AF841293

NONLINEAR DYNAMICS OF INTERACTING POPULATIONS

WORLD SCIENTIFIC SERIES ON NONLINEAR SCIENCE

Editor: Leon O. Chua
University of California, Berkeley

Series A. MONOGRAPHS AND TREATISES

Published Titles

Volume 9: Frequency-Domain Methods for Nonlinear Analysis: Theory and Applications
G. A. Leonov, D. V. Ponomarenko, and V. B. Smirnova

Volume 11: Nonlinear Dynamics of Interacting Populations
A. D. Bazykin

Volume 12: Attractors of Quasiperiodically Forced Systems
T. Kapitaniak and J. Wojewoda

Volume 13: Chaos in Nonlinear Oscillations: Controlling and Synchronization
M. Lakshmanan and K. Murali

Volume 14: Impulsive Differential Equations
A. M. Samoilenko and N. A. Perestyuk

Volume 15: One-Dimensional Cellular Automata
B. Voorhees

Volume 16: Turbulence, Strange Attractors and Chaos
D. Ruelle

Volume 17: The Analysis of Complex Nonlinear Mechanical Systems: A Computer Algebra Assisted Approach
M. Lesser

Volume 19: Continuum Mechanics via Problems and Exercises
Edited by M. E. Eglit and D. H. Hodges

Volume 20: Chaotic Dynamics
C. Mira, L. Gardini, A. Barugola and J.-C. Cathala

Volume 21: Hopf Bifurcation Analysis: A Frequency Domain Approach
G. Chen and J. L. Moiola

Volume 22: Chaos and Complexity in Nonlinear Electronic Circuits
M. J. Ogorzalek

Volume 23: Nonlinear Dynamics in Particle Accelerators
R. Dilão and R. Alves-Pires

Volume 25: Chaotic Dynamics in Hamiltonian Systems
H. Dankowicz

Forthcoming Titles

Volume 4: Methods of Qualitative Theory in Nonlinear Dynamics (Part I)
L. Shilnikov, A. Shilnikov, D. Turaev and L. O. Chua

Volume 18: Wave Propagation in Hydrodynamic Flows
A. L. Fabrikant and Y. A. Stepanyants

Volume 24: From Chaos to Order
G. Chen and X. Dong

Volume 27: Thermomechanics of Nonlinear Irreversible Behaviours
G. A. Maugin

Volume 30: Quasi-Conservative Systems: Cycles, Resonances and Chaos
A. D. Morozov

Volume 31: CNN: A Paradigm for Complexity
L. O. Chua

Volume 32: From Order to Chaos II
L. P. Kadanoff

WORLD SCIENTIFIC SERIES ON NONLINEAR SCIENCE

Series A Vol. 11

Series Editor: Leon O. Chua

NONLINEAR DYNAMICS OF INTERACTING POPULATIONS

Alexander D. Bazykin

Institute of Mathematical Problems in Biology
Russian Academy of Sciences

edited by

Alexander I. Khibnik
Cornell University

Bernd Krauskopf
Vrije Universiteit Amsterdam

World Scientific
Singapore • New Jersey • London • Hong Kong

Published by

World Scientific Publishing Co. Pte. Ltd.
P O Box 128, Farrer Road, Singapore 912805
USA office: Suite 1B, 1060 Main Street, River Edge, NJ 07661
UK office: 57 Shelton Street, Covent Garden, London WC2H 9HE

Library of Congress Cataloging-in-Publication Data
Bazykin, A. D.
Nonlinear dynamics of interacting populations / Alexander D. Bazykin ; edited by Alexander I. Khibnik, Bernd Krauskopf.
p. cm. -- (World Scientific series on non-linear science, Series A ; vol. 22)
Includes bibliographical references.
ISBN 9810216858 (alk. paper)
1. Population biology -- Mathematical models. 2. Biotic communities -- Mathematical models. 3. Bifurcation theory. I. Khibnik, Alexander I. II. Krauskopf, Bernd. III. Title. IV. Series: World Scientific series on nonlinear science. Series A, Monographs and treatises ; v. 22.
QH352.B39 1998
577.8'8'0151--dc21 98-11768
CIP

British Library Cataloguing-in-Publication Data
A catalogue record for this book is available from the British Library.

Copyright © 1998 by World Scientific Publishing Co. Pte. Ltd.

All rights reserved. This book, or parts thereof, may not be reproduced in any form or by any means, electronic or mechanical, including photocopying, recording or any information storage and retrieval system now known or to be invented, without written permission from the Publisher.

For photocopying of material in this volume, please pay a copying fee through the Copyright Clearance Center, Inc., 222 Rosewood Drive, Danvers, MA 01923, USA. In this case permission to photocopy is not required from the publisher.

Printed in Singapore by Uto-Print

Contents

Foreword

This book could be called *Mathematical ecology beyond the Lotka-Volterra model* and we might add: *far beyond.* In theoretical biology and applied mathematics, the logistic and Lotka-Volterra models have long been considered as seminal examples of modeling and dynamics. However, it was understood only recently how different forms of regulatory mechanisms, like birth and death, competition, consumption and the like, result in changes of the stability and dynamics of ecological systems. The present book brings this understanding to the attention of a broad biological and nonlinear dynamics audience. It does so with a deep and unique insight into the mathematical richness of basic ecological models and how this richness emerges as the number of competing mechanisms or factors (reflected in the number of parameters, not state variables) increases.

The main topics of the book are:

- the dynamics of elementary ecological communities, consisting of two or three trophic levels;
- the stabilizing and destabilizing role of various regulatory mechanisms determining the outcome of ecological interaction within the community;
- "dangerous boundaries" for the stability of the ecosystem and criteria for approaching them.
- mechanisms of spatial inhomogeneity and their relationship to non-equilibrium dynamics of ecosystems.

The strength of this book is that it systematically builds a sequence of well-motivated ecological models of increasing difficulty and classifies them with methods from bifurcation theory. To this end, the author emphasizes the use of higher order degeneracies. This makes this book quite unique and interesting not only for a biological audience, but also for the applied dynamical systems community. In fact, this text can be used as a guided tour to bifurcation theory from the applied point of view. The interested reader will find a wealth of intriguing examples of how known bifurcations occur in (biological) applications.

There is a clear structure throughout, and we feel that it will be of help to the reader to sketch it here. All models, especially in the analysis of two-species ecosystems, are put into a matrix-like structure reflecting the interaction between different stabilizing and destabilizing factors. Each model is then reduced by scaling to a convenient form and is analyzed by means of bifurcation theory. A complete description of the bifurcation diagram is given and illustrated in the figures. The phase portraits and bifurcations are then explained in detail with the emphasis on changes in phase or parameter space that lead to qualitatively different behavior. In a last step, rewarding especially for those who are primarily interested in the biological implications, this bifurcation analysis is interpreted from the ecological point of view. Each chapter has an appendix (not included in the Russian edition) containing numerically computed phase portraits together with the equations and parameter values. When compared with the figures in the text, they give an impression of the physical appearance of the systems. The reader is encouraged to investigate the models with any simulation program.

The bulk of the material in this book is based on original research that the author conducted with his collaborators in the 1970s and 1980s at the Institute of Mathematical Problems in Biology in Pushchino, Russia. It was originally published in Russian in 1985. The author worked on an English translation of the original in its revised and extended form, but due to his tragic death in 1994 he could not complete this project. We have assumed the role of translation editors in an effort to finish the project. Our editing consisted essentially in making the text more accessible by using a language of modern bifurcation theory that we consider fairly standard. We tried to keep the spirit of the original as much as possible and included the original preface together with a biographical introduction provided by the author's family.

We thank Elena P. Kryukova (Bazykina), Yegor A. Bazykin and Dmitry A. Bazykin for their continuous support and, in particular, for the preparation of the appendices. For their encouragement and advice on our editing, we thank Faina S. Berezovskaya, John Guckenheimer, Alexey S. Kondrashov and Emmanuil E. Shnol. Finally, we thank the editorial staff at World Scientific for their good cooperation.

Ithaca/Amsterdam — *Alexander Khibnik*
June 1997 — *Bernd Krauskopf*

A. D. Bazykin
(8 June 1940–23 January 1994)

Alexander D. Bazykin*

Dr. Alexander Bazykin first became interested in mathematical biology while studying at the newly created Department of Biophysics at the Physics Faculty of Moscow State University. The department's founding in 1960 was indeed a revolutionary event. Nearly destroyed during the Stalin and Kruschev eras, the rebirth of the science of genetics in Russia in the early 1960's made the department's existence possible. This provided a young Sasha Bazykin with a fascinating "new" field on which to focus his scientific interests. He was always proud to consider himself a student of such brilliant scientists as Nikolai Timofeev-Resovski, Mikhail Bongardt, Alexei Lyapunov, Albert Molchanov and Alexandr Lyubischev. For his senior thesis in 1965, Sasha Bazykin worked on one of the most controversial problems of the theory of microevolution: the possibility of simpatrick speciation. Subsequently, he would spend his life's work in the area of mathematical modeling of populations and ecosystems, becoming one of the founders of that branch of science in Russia.

After completing his undergraduate work, Dr. Bazykin moved to the newly established branch of the USSR Academy of Sciences, Academgorodok, located in western Siberia. Academgorodok's creative and relatively unrestricted experimental environment during this period made it a kind of scientific oasis, attracting both established and promising young scientists from throughout the Soviet Union. Dr. Bazykin worked in the Elementary Mechanisms of Evolution Laboratory founded by Professor Nikolai Vorontsov. There he worked on problems of mathematical modeling of speciation, and wrote and defended his Ph.D. dissertation "Selection and Genetic Divergence in Systems of Local Populations and Populations with Connected Habitat: a Mathematical Model". He also greatly expanded his knowledge of field zoology and ecology by undertaking extensive field work throughout the Asian portion of the former Soviet Union.

At Academgorodok, Dr. Bazykin evolved into a unique expert in the area of mathematical biology. Among his achievements was an in-depth mathematical analysis of genetic polymorphism and, in particular, of the role of stabilizing selection in its formation. Dr. Bazykin was one of the first since the work of Kolmogorov

*Provided by author's family.

and Fisher in the 1930's to return to an analysis of the time-space structure of population processes. In doing so, he proved the possibility of non-correlation between genetic and geomorphologic boundaries, as well as analyzed the problem of partial geographic isolation.

In 1972, Dr. Bazykin was given yet another opportunity to work at an innovative scientific institute. Professor Albert Molchanov established the Research Computing Center (now called the Institute of Mathematical Problems in Biology) at the USSR Academy of Sciences Scientific Center for Biological Research in the town of Pushchino, near Moscow, and invited Dr. Bazykin to join. His move there in 1973 marked the beginning of a tenure that would span twenty-one years. Dr. Bazykin not only founded and headed the Laboratory of Mathematical Problems in Biology for many years, but later also served as the Center's deputy director. In fact, he became one of the primary researchers who shaped the scientific image of the town.

It was during those years at Pushchino that Dr. Bazykin's interests shifted towards the mathematical analysis of elementary ecological communities. He worked out a complete theory of dynamic processes in such communities, paying special attention to qualitative restructuring of dynamic behavior. On the basis of this research, he formulated the concept of dangerous intensity boundaries for dynamic and parameter influences on an ecosystem, and worked out some common criteria for approaching such boundaries. The ideas from this research were summed up in his Doctoral dissertation and in *Mathematical Biophysics of Interacting Populations*, the Russian edition of this book that was first published in 1985.

Beyond his own research, Alexander Bazykin actively participated in many other scientific projects throughout his career, both domestically and internationally. From 1974 on he served as one of the organizers of the annual symposia on mathematical modeling of complex biological systems, headed by Albert Molchanov. From 1975 through 1977 he worked at the International Institute for Applied System Analysis (IIASA) in Vienna as the only Soviet scientist participating in the international interdisciplinary research team, headed by Professor C. S. "Buzz" Holling (currently at the University of Florida). The widely cited monograph resulting from this work, *Adaptive Environmental Assessment and Management* (J. Wiley and Sons, 1978), laid the foundation for the modern concept of ecological monitoring. Following his work at IIASA, he acted as Vice-Chairman of the Section of Mathematical Modeling and System Analysis within the USSR Academy of Sciences Council on the Biosphere. He also served for ten years on the editorial board of the journal *Ecological Modeling*, as well as translated into Russian and edited translations of more than a dozen books on mathematical biology, ecology, and ethology. Finally, in 1990 when Professor Nikolai Vorontsov assumed the post of Minister of Environment in the Gorbachev cabinet, he invited Dr. Bazykin to serve as his Deputy Minister. Dr. Bazykin served in the government for two years, creating and implementing the use of environmentally oriented Geographic Information Systems (GIS), among other projects.

In 1992 Dr. Bazykin began what turned out to be one of his last projects. World Scientific had offered to publish the English edition *Nonlinear Dynamics of Interacting Populations* of his book from 1985 on interacting populations. Extremely enthusiastic about the idea, he started preparing the book for publication when he was diagnosed with a terminal brain tumor. Upon his death in January 1994, the work remained unfinished.

The hard work of updating and preparing this monograph for publication was assumed by Dr. Bazykin's friends, colleagues and students. We are forever indebted to Alexander Khibnik and Bernd Krauskopf for their selfless and often ungratifying work on this edition. It is because of them, of their devotion to science and to the memory of Dr. Alexander Bazykin, that you are holding this book today.

Moscow/Baltimore
May 1997

Elena P. Kryukova (Bazykina)
Yegor A. Bazykin
Dmitry A. Bazykin

The author and his wife, Elena, in 1976.

Preface

The anthropogenic pressure on the environment evidently increases steadily. By all accounts, this development will continue in the foreseeable future. This makes it necessary to minimize the devastating consequences of anthropogenic influences on natural systems. The first step in this direction consists of learning how to estimate the character and size of the impact of these influences and to predict their consequences. A system of ecological monitoring is now being developed both on the global and national scales to meet these objectives (Izrael', 1976, 1977; Izrael' *et al.*, 1981).

Assessing environmental influences and predicting their consequences are closely related problems despite some significant differences. Predicting the consequences of influences would remain a special and very complex task, even if we had reliable "snapshots" of natural systems and knew their dynamics over a certain period of time. There are many possible reactions of ecosystems, none of which are yet studied well enough. However, we can distinguish, although somewhat artificially, the two most important ones (Bazykin, 1978; Holling, 1978).

1. *Buffered reaction of natural systems to external influences.*

For a very wide range of systems, we know that there exists a level of external influences which is called a *threshold* (or *critical level*). Relatively weak influences below threshold, are, in a sense, absorbed by natural systems. The result is only a small, quantitative change, which is undetectable for an external observer in many cases. This ability of ecological systems to withstand, up to a certain extent, external influence is called *resilience* of ecological systems (Holling, 1973). If the intensity of an external influence exceeds the threshold then the system cannot endure the pressure any longer. It breaks down and turns into a qualitatively different and, as a rule, undesirable state.

The significant feature of such a qualitative transition consists of its practical irreversibility. A disappearance of the external influence does not lead to the restoration of the original state of the ecosystem, and it is not possible to get the system back to that state artificially. Only the process of ecological succession, which requires tens and hundreds of years, can contribute to the restoration of the original state.

The distinction between qualitative and quantitative, as well as between gradual and abrupt changes, depends of course on the point of view. In the first place, it is a question of the time required for those changes to occur as compared to characteristic times of the system. In the case of ecological systems, changes must be considered as abrupt (or qualitative) that evolve in the course of several years or decades. On the one hand, this length of time is most relevant for making predictions while, on the other hand, changes that occur over such long periods seem to be gradual from the conventional human point of view. It poses psychological difficulties to perceive such changes as abrupt.

2. *Counter-intuitive reaction of natural systems to external influences.*

The term *counter-intuitiveness* was introduced by J. Forrester (1971), the recognized American expert in the field of systems analysis, as applied to the management of economic, demographic and social processes in large cities, in particular. It refers to reactions of complex systems to external influences that go against common sense. In natural systems they are the rule rather than the exception: excessive application of insecticides leads, in due course, not to the suppression of an insect pest, but to a series of outbreaks; the extermination of predators may lead not to an increase, but to a drop in the numbers of key-industry animals; redundant irrigation entails, in many cases, not improved fertility of agricultural land, but its salinization, etc. The eutrofication of freshwater ponds can serve as a striking example of counter-intuitive consequences of anthropogenic influences. Fertilizers arriving at lakes and storage ponds from fields frequently do not lead, as might be expected, to an increasing productivity of the ponds. Instead they induce a fundamental structural reorganization of the ecological system of the pond resulting in its ecological destruction.

In situations like this, mathematical modeling may be regarded as the most promising tool for predicting the reaction of natural ecosystems to external influences. However, scientists run into serious obstacles when taking this path, obstacles that are intrinsic to any attempt to apply mathematical methods to biology. However, in mathematical models the ecological effects appear, perhaps, in their purest form with all their specific features. Difficulties are mainly the result of two closely related circumstances.

First, the structure of ecological systems is very complicated: they consist of many tens and hundreds of populations of separate species interconnected by thousands of different and, what is particularly important, essentially nonlinear interactions.

Second, all biological systems are unique, and this uniqueness becomes strikingly apparent when we examine natural ecological systems. Therefore, the contradiction between the adequacy and precision of a model on the one hand, and its size on the other hand, is most pronounced when we undertake the mathematical analysis of ecological communities.

This contradiction is clearly reflected by two kinds of methods of mathematical analysis of ecological communities which are, to a considerable extent, independent of each other. The first kind is called *simulation modeling* and aims at achieving the best approximation of a single ecological object, as well as at describing it as concisely as possible. The second kind is called *mathematical modeling* (proper) and tries to describe and to analyse mathematically the characteristics typical for the widest conceivable range of ecological systems (Smith, 1974).

One of the most urgent problems demanding our attention is the stability of ecosystems (see, for example, May, 1974; Svirezhev, Logofet, 1978). Although it is usually considered a property of an ecosystem itself, it would be more correct to regard stability as a property intrinsic to a particular functional regime of an ecosystem. In light of this, there is an urgent need for a systematic study of all models of ecosystems with more than one possible locally stable functional regime. Such models are most adequate for describing qualitative changes of regimes of ecosystems under external influences.

Here it is natural to consider two classes of phenomena (Bazykin, 1982). The first class consists of qualitative changes of the functional regime as a result of a single, non-permanent external disturbance. Such an influence can move the ecosystem away from one stable regime and to another, qualitatively different one. The second class consists of qualitative changes that occur under the influence of a constant external change of gradually increasing intensity. When this intensity exceeds a certain threshold, the basin of attraction of a regime may for instance shrink to a point and disappear. As a result, the ecosystem reorganizes itself, which is practically irreversible: it changes its functional regime. Adequate mathematical tools to analyse such changes are the qualitative theory of differential equations and bifurcation theory (Andronov *et al.*, 1971, 1973; Arnol'd, 1983).

The purpose of this book is the systematic analysis of different dynamic regimes in models of two or three interacting populations that are interconnected by different biological interrelationships. Primary attention is given to the nonlinear dynamical effects in the modeled systems, depending on their initial state and the external conditions. They allow for different possible functional regimes, as well as for qualitative restructuring under the influence of external factors.

A careful, systematic study of greatly simplified interaction models of only two or three populations may be of interest for theoretical and practical ecology for the following reasons:

1. There are serious reasons to believe that the fundamental behavioral characteristics of a natural ecosystem (such as the above-mentioned buffered and counter-intuitive reactions to external influences) are due not so much to its complexity (number of components or species), but to the pronounced nonlinear character of the relationships between separate components of the

ecosystem. This may be observed even in a model ecosystem consisting of only two or three components. Analysing and classifying counter-intuitive behavior and criteria for when a threshold of external influences is approached for simple models is also of use for real, incomparably more complex ecosystems.

2. The near-threshold behavior of a complex system may be well described by the respective approximating system consisting of few variables (Molchanov, 1975*a,b*).
3. The study of interacting populations consisting of only two or three species extracted from an ecosystem seems to be ecologically justified as well as of practical interest in a number of important cases. Systems, as "forest–pest", "agricultural crop–agricultural pest", "valuable marketable animal species–its main resource–its major predator", and so on, serve as examples.

The largerly nonlinear dynamics of interacting populations require using both analytical and numerical methods. The analytical study allows us to apply results from the qualitative theory of differential equations and bifurcation theory (Arnol'd, Il'yashenko, 1988). These are essentially used for analysing the bifurcations of codimension one, two and three that occur in population models. A numerical study was carried out by applying two software packages (Levitin 1989; Khibnik, 1990; Khibnik *et al.*, 1993) developed at the Institute of Mathematical Problems of Biology (formerly Research Computing Centre), Pushchino: the TRAX package for analysing the phase space of dynamical systems depending upon parameters, and the LOCBIF package for constructing their bifurcation diagrams.

Some results of the numerical analysis are presented in the appendices after the individual chapters. We would like to draw special attention to the correspondence between the figures appearing in the book itself and the computer pictures presented in the appendices. In our opinion, both of them complement each other, giving a new understanding of dynamical models. Moreover, it seems to us that the behavior of strongly relaxational systems cannot be understood without an analytical prediction of the phase portraits which, in a sense, serves as a guide. Furthermore, the numerical study of systems gives an insight into dynamical characteristics which could hardly be discovered within the framework of analytical studies.

Finally, it is necessary to emphasize that this book could never have appeared without the enormous assistance of my closest collaborator, Faina S. Beresovskaya. I can never hope to repay her for her support. The present work has been greatly influenced by long-time scientific collaboration with researchers of the Institute of Mathematical Problems in Biology, (formerly R. A. S.) especially with A. I. Khibnik, Yu. A. Kuznetsov, E. A. Aponina and Yu. M. Aponin, as well as with my colleagues from the Ecological Centre, Pushchino. The computer-generated figures of the book were prepared with the assistance of S. L. Zudin who spent a lot of his time on that. This book was translated into English with great consideration and diligence by Vladimir V. Ievenko. The immense work of producing the camera-ready manuscript was done by P. Ya. Grabarnik, whom my younger son George Bazykin assisted.[1] I am particularly grateful to all of them. The generous support of *the George Soros Foundation* significantly contributed to this study, enabling the author to explore new computer technologies.

[1]This refers to a previous version of the book.

Chapter 1

Ideas and Methods of Modeling Populations

Arguably, the history of the application of mathematics to ecology dates back to the book *An essay on the principle of population* by Malthus (1798). There it is mentioned for the first time that a population with an opportunity to reproduce grows exponentially in time. In modern notation and terms, the dynamics of a population with no resource limitations can be described by the equation

$$\dot{x} = ax\,. \tag{1.0.1}$$

This is known as the exponential growth equation, since it has the solution $x(t) = x_0 e^{at}$. Certainly Malthus had predecessors, beginning with the Italian mathematician Fibonacci in the twelfth and thirteenth century, who is credited for the well-known problem of how many pairs of rabbits will be born to single pair of rabbits year after year. However, it is Malthus who deserves the credit for stating the universal law of population growth in a clear and unambiguous way. We will not dwell here on the economic and political views of Malthus which have been severely criticized in the literature. The only relevant thing for us is the highly idealized concept of a completely homogeneous population, in which the individuals are identical and population growth is unlimited. This notion turned out to be as fundamental for the development of mathematical ecology as the idealized concept of a dimensionless point of mass for the development of mechanics.

The next step in the field was introducing a model of a population that is restricted in size by some necessary but limited resource. Verhülst (1838) described the dynamics of such a population by the equation

$$\dot{x} = ax(K - x)/K\,, \tag{1.0.2}$$

which has since become known as the logistic equation. Here, a is the rate of exponential population growth at smaller population size, and K is the stationary population density, determined by the available resources. Later this work was forgotten. After Pearl (1927; 1930) rediscovered equation (1.0.2), it has been known in the ecological literature as the Verhülst-Pearl equation.

The contributions mentioned above were intended to describe the dynamics of a single population, primarily the human population. The first mathematical-ecological studies that truly aimed at describing interacting populations appeared as late as the 1920's (Lotka, 1925, 1956; Volterra, 1926, 1931).

Their most important impact was to demonstrate how purely mathematical methods can lead to conclusions about the dynamics of a system on the basis of only a few biologically plausible and experimentally verifiable assumptions about inter- and intraspecies interaction. The best known conclusion concerned the possibility of endogenic fluctuations in the sizes of two populations interacting as a predator-prey system. Gause and his coauthors (Gause, 1933; Gause, Vitt, 1934; Gause, 1934) worked on the experimental verification of the results obtained by Volterra and Lotka and developed some mathematical principles to validate their studies. Unfortunately, their work was interrupted too early.

The studies of Volterra also initiated the work of Kolmogorov (Kolmogorov, 1936, 1972) who suggested a conceptually new approach to the problems of mathematical ecology: assumptions about the nature of inter- and intraspecies interactions should be formulated without explicitly specifying functional dependencies, which cannot be found experimentally. Instead, Kolmogorov maintained that they should be modeled only by specifying the *qualitative* features of the corresponding functions. It was shown that even in that case, mathematical techniques do provide substantial biological conclusions about the nature of the dynamics. In the late thirties, this pioneering stage in the development of mathematical ecology ended. It can be said that, although isolated work continued, a long pause ensued in the overall development of this field.

The begining of a new stage of intense development of mathematical ecology, which is continuing even now, came in the 1960's and was due to two circumstances. First, the catastrophic consequences of the antropogenic impact on natural ecosystems had added to the urgency of predicting these impacts. One of the most effective methods for this problem seemed, and still seems, to construct and analyse mathematical models of the systems under investigation. Second, rapid progress in computing and the successful use of computers for solving problems in a variety of fields had lead to the natural hope that they could also be applied to problems in ecology. This technological progress resulted in an intense development of simulation modeling (Moiseev, 1979).

The merits of simulation modeling are obvious: In a number of cases, the construction and implementation of models of ecosystems yields reliable predictions of their dynamics. These sometimes even lead to accurate predictions of the reaction of an ecosystem to external influences (e.g., Menshutkin, 1971; Zhdanov, Gorstko, 1975; Gorstko, 1976; Skaletskaya *et al.*, 1979). However, the possibilities of this method are limited, primarily due to difficulties in determining the range of application of a simulation model. In particular, the period of time for which predictions can be made with a desired accuracy may be unclear. Furthermore, a simulation model is, by its nature, always anchored to a concrete object of study, and any attempt to use it for another, even a related, object calls for a significant modification of the model. Finally, simulation models are meant to be used to model comparably small fluctuations in ecosystems with relatively small variations in the living conditions. In practice we are often interested in understanding drastic changes in an ecosystem's dynamics resulting from small or large changes in the environmental conditions.

These limitations of simulation modeling have a common reason. The construction and numerical computation of an *exact* model can only be successful in areas where there is an exact quantitative theory. That means, there are equations to describe a given phenomena, and the task consists of solving these equations with a prescibed accuracy. If an appropriate quantitative theory is not available, constructing an exact model is of limited value.

The realization that simulation modeling are limited caused a group of ecologists to replace their initial enthusiasm with reasonable skepticism (Holling, ed., 1978; Molchanov, Bazykin, 1979). This revived an interest in mathematical modeling itself, which developed as a separate field that had little to do with simulation modeling until recently. A tendency to combine these two fields has only been observed in the last few years.

We should mention that mathematical modeling in ecology, or mathematical biophysics of populations and communities, has not yet reached the status of a separate scientific field. Many of the recent monographs describe the use of various mathematical techniques either to treat a specific (often quite general) biological problem (e.g., May, 1974; Svirezhev, Logofet, 1978), or to analyse various ecological systems (Pykh, 1983; Shapiro, Luppov, 1983), although there are naturally studies of an intermediate nature (Poluektov, ed., 1974). In this introductory chapter, we therefore want to mention the main and, in our opinion, the most interesting trends of research that are closely related to the present work. We also indicate the place that the subject matter of this book occupies among these trends. However, we do not pursue the global task of analysing the current state of mathematical biophysics of populations and communities.

Constructing a mathematical model of any object or phenomenon inevitably demands some degree of idealization. The logic of mathematical modeling is such that the more idealized and simplified concepts we use, the more general are the properties of the studied objects that can be analysed. As Romanovsky remarked, maximally simplifying a model and decreasing the number of independent variables, however paradoxical it may be, leads to a deeper understanding of the modeled phenomenon (Romanovsky *et al.*, 1975). On the other hand, for understanding different aspects of a single phenomenon, various idealizations of the same object may be necessary. We list here those assumptions that are widely used in mathematical biophysics of populations and communities, and relate them to suitable mathematical techniques and biological problems.

1. In the overwhelming majority of publications on mathematical ecology, the external conditions are assumed to be constant, because it is quite natural to analyse the properties of an autonomous system prior to studying the role of external effects. This gives rise to models described by differential or difference equations with constant coefficients. Nonetheless, interesting attempts have been made to estimate the effect of small fluctuations of external conditions on the ecosystem dynamics (Freidlin, Svetlosanov, 1976; Sidorin, 1981). Of particular interest here are the situations in which a system has several attractors in the absence of pertubations. However, this leads to serious mathematical difficulties, so that results have only been obtained for the simplest case of an isolated population with several equilibria.
2. As a rule, natural populations consist of hundreds, thousands and sometimes millions or more individuals. When considering very large populations, it is accepted to make use of two idealizations: (1) the population size is described by a continuous value; (2) random fluctuations in population size can be neglected, so that only the dynamics of the average sizes need to be studied. Allowing for random fluctuations requires the use of mathematical techniques from probability theory and the theory of random processes (Moran, 1962; Gorban' *et al.*, 1982). Neglecting these fluctuations leads to the use of deterministic differential or difference equations. A. A. Lyapunov, a pioneer of mathematical modeling in Russia, suggested that the dynamics of an individual population should be analysed by applying stochastic processes, whereas the dynamics of several interacting populations should be studied by means of differential equations (Lyapunov, 1972; Lyapunov, Bagrinovskaya, 1975). Actually, it is methodologically reasonable to neglect fluctuations in the early stages of modeling, and to take into account the additional effects of random fluctuations only in later stages. In so doing, we should estimate characteristic time intervals, for which the consideration of random fluctuations significantly changes the picture.
3. It is common practice in mathematical ecology to use various idealizations for assumptions concerning the age distribution of populations. One of them is that all individuals reproduce synchronously once they reach a certain age. Such an idealization gives rise to difference equations. They were applied to the problems of mathematical ecology for the first time by Leslie (Leslie, 1945, 1948) in order to study the dynamics of the age structure of an isolated population. Later, difference equations were successfully used to analyse the dynamics of separate populations. In particular, they were applied to harvested species with strongly pronounced seasonal fluctuations in breeding (Ricker, 1954). Chaotic fluctuations of the population size were first observed in mathematical ecology, under the assumption of constant external conditions for

populations with discrete non-overlapping generations (Shapiro, 1974; May, 1975). Later, such chaotic dynamic regimes were also found in models of ecological systems with continuous time. As a rule, difference equations are used to analyse changes in the sizes of individual populations. A series of articles by A. P. Shapiro and his colleagues (Shapiro, Luppov, 1983) dealing with the dynamics of two-species communities may be regarded as a certain exception.

The second widely used idealization concerning the age structure of populations is the assumption that generations do overlap, but that the rate of variation of its size is determined by the population size at some previous time. This can be described by delayed differential equations, as proposed for the first time by Hutchinson (1948). The main concerns when using this technique are the existence and characterization of oscillatory behavior. Recently, Yu. S. Kolesov and his coauthors (Kolesov, 1979; Kolesov, Shvitra, 1979*a,b*) completed a large series of articles which used delay equations to analyse the dynamics of systems with two interacting populations. They are still working on this topic and devote much attention to the interesting biological and mathematical problems which arise when the system coefficients satisfy a resonance condition, or are near resonance.

4. Up to now we have considered the idealizations related to the dynamics of an isolated population, or of very few interacting populations. Natural biogeocenoses consist of populations of several tens or hundreds of species. That is why researchers have made repeated attempts to approach the dynamics of such a complex system by applying ideas and methods of statistical mechanics (Kerner, 1955, 1957; Alekseev *et al.*, 1969; Polishchuk *et al.*, 1969; Polishchuk, 1971; Alekseev, 1975). However, owing to the insurmountable mathematical difficulties involved in the development of techniques for the statistical mechanics of nonlinearly interacting particles, very strong idealizations are needed to apply these techniques to the dynamics of biogeocenoses. This, first of all, concerns the postulate known as Volterra's principle of equivalents, as well as the assumption of strictly bilinear interaction between species.
5. All of the idealizations considered so far apply to systems with complete mixing, called *local systems.* In terms of popular biology, this means that an individual, during its lifetime, should have the possibility to be everywhere in the territory inhabited by the population. This is obviously a very strong condition, because the size of the habitat may, in reality, exceed the area an individual can cover in its lifetime by a factor of ten, a hundred or even a thousand. Although temporal dynamics are of exclusive interest while studying models of local, or concentrated, communities, models of spatially distributed communities are studied in both temporal and spatial respect. As a rule, this is done by using diffusion equations with nonlinear right-hand sides, or, using presently accepted terminology, systems of

diffusion-kinetics type (Haken, 1978). Models of spatially distributed communities are much less studied than local models. In fact, work in this field has just begun, and there is currently no sufficiently complete classification of the behavior such models can display. The effects attributed to travelling waves (fronts) (Kolmogorov *et al.*, 1937) and stationary dissipative structures are now receiving much attention (Bazykin *et al.*, 1980; Bazykin, Khibnik, 1982; Razzhevaikin, 1981*a,b*).

This monograph considers communities exposed to constant environmental influences and consisting of two or three interacting populations which are large enough to neglect fluctuations that might be present. The rates at which the population sizes vary are determined by instantaneous values of these sizes, with no consideration of the age structure of the population. These idealizations make it possible to exclusively use ordinary differential equations with constant coefficients and without delay as models. They can be analysed using the qualitative theory of differential equations and bifurcation theory.

What biologically interesting questions arise in the study of models of ecological communities within the framework of the conventional idealizations we have enumerated? We only list the most important of them here:

1. How does a community behave when it is left to itself? What regimes can be established: stationary, oscillatory or chaotic?
2. How does the behavior of such a community depend on its initial state, if it does at all?
3. How does an ecosystem react to environmental influences? What is the effect of a single disturbace of the state of the system (meaning that the corresponding point is phase space is perturbed to another place, after which the system is left to itself.)? What is the effect of a permanent influence (meaning a change of the parameters of the system)?
4. How does the incorporation of spatial inhomogeneity effect the temporal dynamics of an ecosystem and lead to spatio-temporal organization?

Chapter 2

Dynamics of Isolated Populations

In this chapter we study models for the growth of a *single* population. Some of these models will be used in the sequel as constituents for models of *interacting* populations. Furthermore, we introduce the important concept of *dangerous boundaries*, both in phase and in parameter space, and illustrate this with examples. We consider models for

1. a population growing without any limitation, called a free population,
2. a population whose growth is restricted by external resources, and
3. a harvested population.

2.1. Free Population

In this chapter and throughout the remainder of this book, we intend to neglect ages, genotypes and other structural organizations and consider *homogeneous* populations. With the exception of the last chapter, we only study concentrated or local populations for which the concepts of size and density are identical. Under these assumptions of homogeneity, the equation describing the dynamics of a population can be written in the general form of

$$\dot{x} = F(x)\,, \tag{2.1.1}$$

or

$$\dot{x} = xf(x)\,. \tag{2.1.2}$$

Population dynamics involves two processes: reproduction and death of individuals. When we examine those separately, Eqs. (2.1.1) and (2.1.2) take the form

$$\dot{x} = B(x) - D(x)\,, \tag{2.1.3}$$

and

$$\dot{x} = x[b(x) - d(x)]\,, \tag{2.1.4}$$

respectively. Here, $B(x)$ and $D(x)$ are the absolute reproduction and mortality rates of individuals, and $b(x)$ and $d(x)$ are the corresponding per capita reproduction and death rates, that is, fecundity and mortality in common terminology.

Now we consider what form the functions $b(x)$ and $d(x)$ might take, and we qualitatively analyse the solutions of Eq. (2.1.4).

When studying a free population, it is customary to assume that the mortality is independent of the population size. Sometimes it is reasonable to assume that the mortality rate is negligible, or even to consider the highly idealized case that $d(x) \equiv d = 0$.

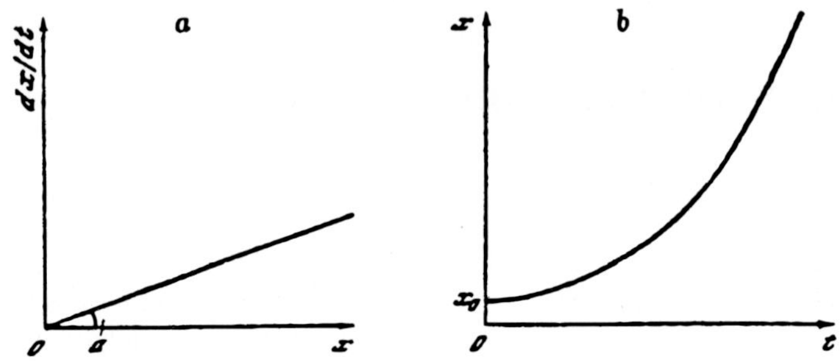

Fig. 2.1.1. Two graphical representations of the dynamics of Eq. (2.1.5); x_0 is the initial size.

With fecundity it is somewhat more complicated. In the simplest case, fecundity and mortality are both independent of the population size. Under this assumption we arrive at the classical equation of Malthus, the exponential growth equation,

$$\dot{x} = (b - d)x = ax\,. \tag{2.1.5}$$

Note that for $b < d$ this equation describes the exponential decay of the population. The dynamics described by Eq. (2.1.5) is shown in Fig. 2.1.1.

The assumption that fecundity is independent of the size of the population is best suited in the context of asexual reproduction, for instance through mitosis, by which the reproduction of an individual does not depend on the presence of other individuals.

Let us now examine a population that grows by sexual reproduction. In this case it is natural to expect that the absolute reproduction rate $B(x)$ should be proportional to the frequency of contacts between individuals. If we assume that individuals move in the population like Brownian particles then this frequency is proportional to the squared population density. Since density and size are synonymous for the local models considered here, one gets

$$B(x) = xb(x) = kx^2\,. \tag{2.1.6}$$

Hence, under the assumption of negligible mortality this gives

$$\dot{x} = kx^2\,.$$

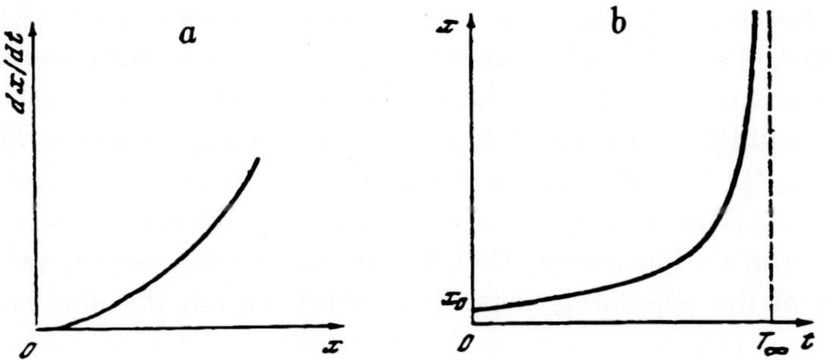

Fig. 2.1.2. Two graphical representations of the dynamics of Eq. (2.1.5).

The solution to this equation is a hyperbola having a vertical asymptote

$$x(t) = x_0 T_\infty/(T_\infty - t); \qquad T_\infty = 1/kx_0\,. \tag{2.1.8}$$

In other words, not only does the size of a population governed by (2.1.6) grow without bound, but it even exceeds any bound at the moment T_∞ (see Fig. 2.1.2). It is clear that this is only realistic for low population densities. However, it is interesting to note that the human population growth, starting from the period for which reliable estimates are available up to about the end of the 1960's, is in very good agreement with the dependence given by (2.1.8) (Foerster *et al.*, 1960; Shklovsky, 1965; Watt, 1968). This gives some evidence that the reproduction law (2.1.6) may arise not only due to random contacts among the individuals, but also because of some other mechanisms yet to be explored. Effects like cooperation among the individuals of colonial animals (MacFadyen, 1963) are among these alternative mechanisms.

Obeying the reproduction law in (2.1.6) and allowing natural mortality to be independent of the population density results in the following equation for the population dynamics:

$$\dot{x} = kx^2 - dx = x(kx - d)\,. \tag{2.1.9}$$

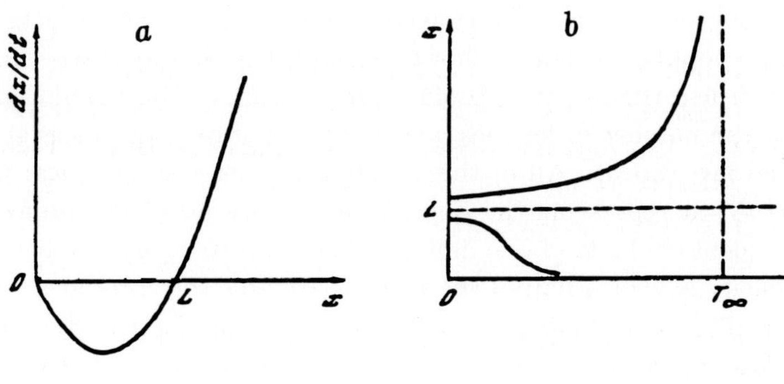

Fig. 2.1.3. The dynamics of Eq. (2.1.9)

At large population sizes, the dynamics of a population obeying Eq. (2.1.9) does not differ from that of a population satisfying Eq. (2.1.7): the size exceeds any bound after some finite time (Fig. 2.1.3 and 2.1.4). At small sizes, however, the behavior of the population differs qualitatively from the ones considered above. Figures 2.1.3*a* and *b* show that the rate of change in the population size is subject to sign inversion depending on the population size. In particular, this rate is negative if $x < d/k$ and positive if $x > d/k$. This means that, in contrast to the cases considered above, the behavior of the population depends qualitatively on the initial condition: if the initial population is larger than the threshold $L = d/k$ the population grows without bound, and it dies out otherwise.

The concept of a threshold of the population size or density has a natural biological interpretation. A density below threshold is so small that, figuratively speaking, individuals die out more frequently than they meet each other. To be more precise, the threshold of the population density is reached if the average time between subsequent contacts of potential breeding partners is equal to the mean lifetime of an individual, divided by the average number of offspring from reproduction.

The behavior of populations obeying Eqs. (2.1.7) and (2.1.9) is obviously nonbiological at large sizes, since not only the population size, but also the relative rate of population growth exceeds any bound after some finite time. It seems more reasonable biologically to suppose that fecundity depends on the population size. We can characterize this relationship with the following compromise between Eqs. (2.1.5) and (2.1.6) (Bazykin, 1969):

$$b(x) = bx/(N + x)\,. \tag{2.1.10}$$

Within the framework of this model, the population grows according to the hyperbolic law (2.1.6) at small sizes $x \ll N$, and according to the exponential law (2.1.5) at large sizes $x \gg N$.

The dependence in (2.1.10) has a natural interpretation. It should be noted that in a sexually reproductive population, a female which is fertilized does not participate in the reproduction process for some time τ. As a result, at low population density when the characteristic time between the contacts of individuals is much more than τ, the growth rate is determined by the frequency of contacts and obeys (2.1.6). At large population density, however, when the characteristic time between contacts is much less than τ, the absolute growth rate of the population is determined only by the number of females, and the population reproduces according to the exponential law (2.1.5). All of these facts elucidate the biological meaning of the parameter N: it represents the population density at which the average time between subsequent contacts of one individual is equal to τ, or, in other words, the population density at which half of the females are able to reproduce.

When mortality is negligible ($d = 0$), the equation for the dynamics of the population takes the form

$$\dot{x} = bx^2/(N + x)\,. \tag{2.1.11}$$

In this case, the dynamics is slightly different from the exponential, Malthusian dynamics. This difference is only noticeable after we represent the relation between population size and time on a logarithmic scale.

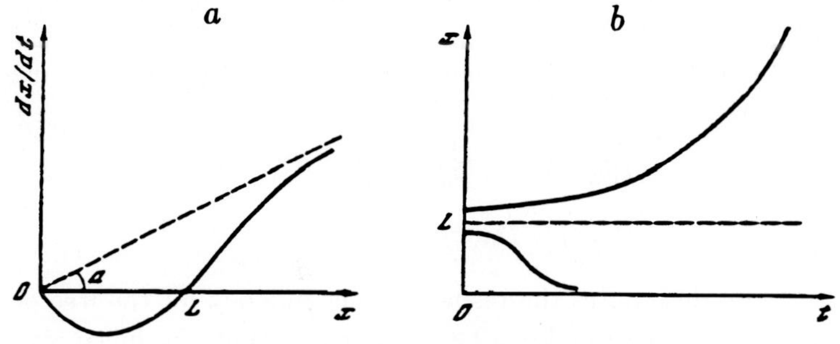

Fig. 2.1.4. The dynamics of Eq. (2.1.13).

Maintaining (2.1.10) as the reproduction law and, at the same time, assuming that mortality is different from zero but still independent of the population size, we arrive at the following equation for the population size:

$$\dot{x} = bx^2/(N + x) - dx\,. \tag{2.1.12}$$

Substituting $b - d = a$ and $dN/(b - d) = L$, we can rewrite this as

$$\dot{x} = ax(x - L)/(N + x)\,, \tag{2.1.13}$$

where it is assumed that $b > d$. As one can see from Fig. 2.1.4, the dynamics described by Eq. (2.1.13) is characterized by the following property. There is a threshold L of the population. If the initial population size x_0 is below this value ($x_0 < L$) then the population dies out. However, if $x_0 > L$ then the population grows without bound, where the growth law becomes exponential as the population grows.

2.2. Population with External Resources

It is obvious that the unlimited growth of populations mentioned above is not possible because of the limitations of external resources such as food, habitats, and so on. The limitation of external resources leads to intraspecies competition, which manifests itself as the dependence of fecundity, mortality, or both, on the population size. As a result, fecundity drops with an increasing population density, whereas the death rate grows. The most simple assumption, generally accepted and confirmed in many cases by experiment, is that these dependencies are linear functions. We denote the dependency of fecundity and mortality on the density of the free population by $b_0(x)$ and $d_0(x)$ respectively, and use the same letters without indices to denote the analogous functions when there are limited resources.

We then obtain

$$\begin{aligned} b(x) &= b_0(x) - e_b x\,, \\ d(x) &= d_0(x) + e_d x\,, \end{aligned} \tag{2.2.1}$$

where e_b and e_d are coefficients of proportionality that show how the competition affects fecundity and mortality.

The general equation for the dynamics of a population which is restricted by resources can be written as

$$\begin{aligned} \dot{x} &= x[b_0(x) - d_0(x) - ex] \\ &= B_0(x) - D_0(x) - \mathrm{ex}^2 = F_0(x) - ex^2\,. \end{aligned} \tag{2.2.2}$$

Here, an index zero designates the respective dependency for the free population.

The quadratic term ex^2 in Eq. (2.2.2) indicates that the linear dependence of fecundity and/or mortality on the population density has another quite natural, though mechanistic, interpretation of intraspecies competition: the absolute shortage of individuals under conditions of limited resources, as compared to the growth of a free population, is proportional to the number of contacts (encounters) between individuals.

Let us explore the consequences of taking into account the limitation of resources, or intraspecies competition, in the models of the previous section.

Equation (2.1.5) becomes

$$\dot{x} = ax - ex^2 = ax(K - x)/K; \qquad K = a/e\,, \tag{2.2.3}$$

which is the famous Verhülst–Pearl logistic equation (Fig. 2.2.1). For any initial state, the population size converges to K monotonically. If the initial size is less than $K/2$, then the graph of size versus time has an inflection at height $K/2$.

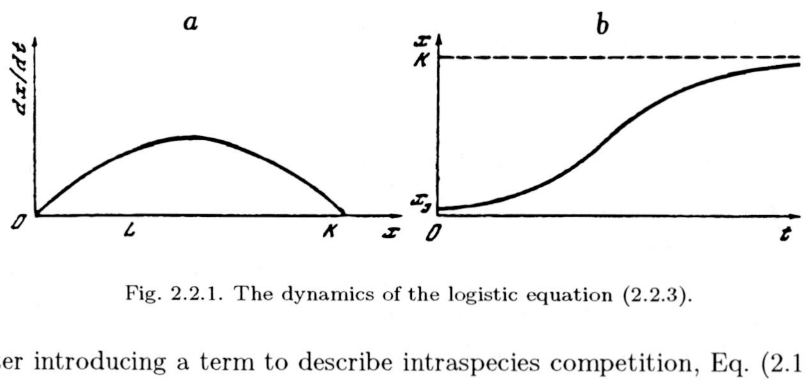

Fig. 2.2.1. The dynamics of the logistic equation (2.2.3).

After introducing a term to describe intraspecies competition, Eq. (2.1.7) does not change its form:

$$\dot{x} = k_0 x^2 - ex^2 = kx^2; \qquad k = k_0 - e > 0\,. \tag{2.2.4}$$

As before, it describes the hyperbolic growth of the population size (see Fig. 2.1.2*b*).

Analogously, Eq. (2.1.9) also retains its form

$$\dot{x} = k_0 x^2 - d_0 x - e x^2 = k x^2 - d_0 x; \qquad k = k_0 - e\,. \tag{2.2.5}$$

The ecological interpretation is obvious: within the framework of the assumptions made to obtain (2.2.4) and (2.2.5), both growth and decrease in the population size occur due to the same factors, namely random contacts between individuals. In other words, the linear dependence of fecundity and/or mortality on the population size is not enough to stabilize the population. Once again, this suggests that models (2.1.7) and (2.1.9) are biologically meaningless for large populations.

Taking intraspecies competition into account, Eq. (2.1.11) can be written as

$$\dot{x} = \frac{b_0 x^2}{N + x} - e x^2 = \frac{b x^2}{N + x} \cdot \frac{K - x}{K}\,, \tag{2.2.6}$$

where $b = b_0 - eN$ and $K = b_0/e - N$.

Qualitatively, the dynamics of a population obeying Eq. (2.2.6) is the same as the dynamics of a population described by the logistic equation (see Fig. 2.2.1*b*). The difference is of a quantitative nature and consists, first, of a slower increase in the population at small population sizes, and, second of the fact that the inflection point on the size-versus-time graph lies above $K/2$. For $N \gg K$ and population sizes less than K, Eq. (2.2.6) can be approximated by the simpler equation:

$$\dot{x} = b x^2 (K - x)/K\,. \tag{2.2.7}$$

Within the framework of model (2.1.13), incorporating intraspecies competition leads to the equation

$$\dot{x} = \frac{b_0 x^2}{N + x} - d_0 x - e x^2 = \frac{a x (x - L)(K - x)}{(N + x) K}\,, \tag{2.2.8}$$

where $a = eK$, and L and K are the roots of the equation

$$x^2 - [(b_0 - d_0)/e - N]x + d_0 N/e = 0\,.$$

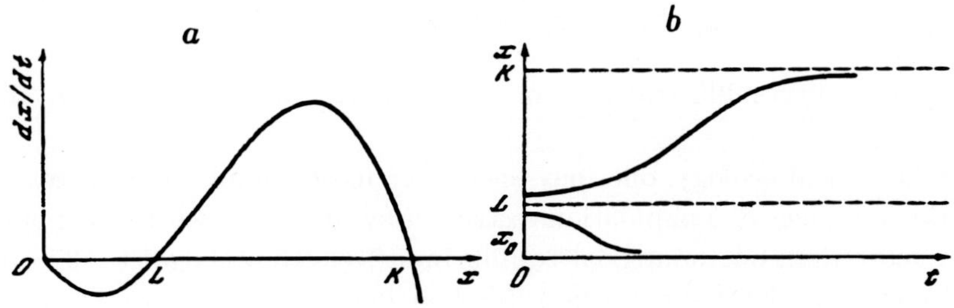

Fig. 2.2.2. The dynamics of the logistic equation (2.2.8).

The corresponding graphs are shown in Fig. 2.2.2. It can be seen that the population described by (2.2.8) has the two nontrivial (that is, nonzero) equilibria

$x = L$ and $x = K$. If the initial population size x_0 is greater than L, then the population increases monotonically, converging to a value $x = K$, just as it does in the case of a population that obeys the logistic equation (2.2.3). If $x_0 < L$ the population dies out. At $x = K$ the population is in a stable equilibrium, whereas the equilibrium at $x = L$ is unstable.

Thus, if a population that was initially in a stable equilibrium $x = K$, falls to a level below L as a result of a single disturbance, then this population is doomed to die out. The unstable equilibrium $x = L$ is the simplest example of a dangerous boundary in phase space.

Let us assume that a population is initially at the equilibrium $x = K$ and is repeatedly subjected to single disturbances that decrease its size. In this case, after each event, the population is allowed to restore its size, but the intensity of the perturbation is increased every time which brings the perturbed population closer to the threshold ($x = L$). How can we know when the intensity of the influence will, indeed, bring the population to a dangerous level? There is a well-known simple ecological criterion (Watt, 1968) for this situation that becomes clear from Fig. 2.2.2*a*: the nearer the perturbation brings the population to the threshold, the slower the population leaves the perturbed state.

Equation (2.2.8) is one of the most simple and natural forms for presenting the Allee effect (Allee *et al.*, 1949; Odum, 1971), which states that fecundity depends non-monotonically on the population size. Actually, in view of the reasons analysed earlier and accepted within the framework of model (2.2.8), the dependence of fecundity on the population size within this model is non-monotonic. If we assume, as earlier, that mortality is independent of the population size, then the points of intersection of the graph of $b(x)$ with the horizontal line d correspond to the unstable ($x = L$) and stable ($x = K$) equilibria of the population.

The Allee effect is formulated in qualitative terms (monotonicity versus non-monotonicity), and the corresponding dependencies can be described by various functions. In particular, if $N \gg K$ then Eq. (2.2.8) is well approximated for the range of population sizes near K by the simpler equation:

$$\dot{x} = ax(x - L)(K - x)\,. \qquad (2.2.9)$$

The respective curves illustrating Eq. (2.2.9) agree qualitatively with those in Fig. 2.2.2.

In mathematical ecology, one may encounter more complicated models for describing the dynamics of a population restricted by resources, where the population may have more than one nontrivial equilibrium (Svirezhev, Logofet, 1978; Huberman, 1978; Denisov, Kuznetsov, 1981). Such models are not discussed in this book.

2.3. Harvested Population

Let us consider a population from which some individuals are regularly removed, a harvested population. Here, we assume that the harvesting intensity remains constant over a significant period of time, although it can take different values in general. In this case, the variation in population size is expressed by

$$\dot{x} = F(x) - S(x, \alpha), \tag{2.3.1}$$

where the function $F(x)$ describes the dynamics of the unharvested population, and $S(x, \alpha)$ is the rate at which the individuals are removed from the population. We call the parameter α the harvesting intensity and note that it can have different meanings depending on the nature of the function $S(x, \alpha)$. A population harvested at an intensity that is piecewise continuous in time is the simplest example of a permanent external influence on the ecosystem.

Now we consider a population whose dynamics obeys the logistic equation (2.2.3) in the absence of harvesting. Let us investigate how two different harvesting strategies affect the dynamics. The first strategy is to remove per time unit a constant number of individuals from the population. The second strategy is to remove per time unit a constant fraction of individuals, that is, a number of individuals proportional to the population size.

The dynamics for the first strategy is described by

$$\dot{x} = ax(K - x)/K - \alpha, \tag{2.3.2}$$

and for the second by

$$\dot{x} = ax(K - x)/K - \alpha x, \tag{2.3.3}$$

where α is the harvesting intensity. For the first strategy it represents the number of individuals, and for the second the fraction of the population removed per time unit.

In both cases, we consider how the population depends on the harvesting intensity. It is convenient to draw the graphs of $F(x)$ and $S(x)$ in the same plot. The abscisses of their intersection points correspond to (stable and unstable) equilibria of the populations harvested under given conditions, while the ordinates correspond to the number of individuals removed from the population per time unit, which is the *yield*.

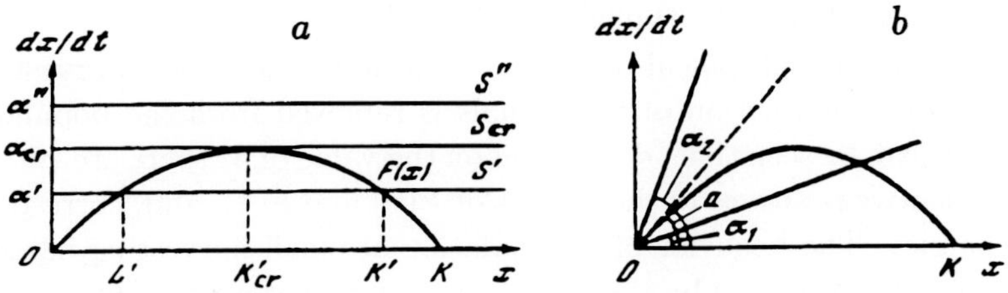

Fig. 2.3.1. Dynamics obeying the logistic law – (a) the first harvesting strategy at a harvesting intensity below (α'), at (α_{cr}), and above (α'') threshold; (b) the second harvesting strategy.

For the first harvesting strategy, Fig. 2.3.1 *a* gives an idea of three possible relative positions of the graphs of $F(x)$ and $S(x)$ for different values of the parameter α.

It can be seen that Eq. (2.3.2) describes two qualitatively different types of dynamics depending on α. When the harvesting intensity is small ($\alpha < \alpha_{cr} = aK/4$) the population has two equilibria, $K'(\alpha)$ and $L'(\alpha)$, where the larger $K'(\alpha)$ is stable and the smaller $L'(\alpha)$ is unstable. The behavior of the population agrees qualitatively with that described by Eq. (2.2.9): there exists a lower threshold of the population size such that the population is doomed to die out if the initial size is less than the threshold. If the harvesting intensity is large ($\alpha > \alpha_{cr} = aK/4$) the population has no equilibria. The population cannot withstand harvesting and is doomed to die out from any initial size. The case $\alpha = aK/4$ is a threshold, at which the stable and the unstable equilibrium merge into the semi-stable equilibrium K'_{cr}.

In order to interpret the results ecologically, imagine that the population is not harvested and is at the stable equilibrium $x' = K$. Let us consider the dependence of the stable population size on the parameter α. As the harvesting intensity increases, the stable equilibrium size K' decreases monotonically to the threshold K'_{cr} and then suddenly drops to zero. In this case, the threshold of the harvesting intensity S_{cr} equals the maximum growth rate of the unharvested population.

Assume now that this population is harvested at a rate which increases from time to time by a finite value, and that after each time the intensity increases, where the population is given sufficient time to reach a new equilibrium. What is the criterion for the harvesting intensity being close to the threshold? Such a criterion is less evident than in the above-mentioned case of single loads, although it is rather obvious from the mathematical point of view. According to this criterion, the closer the harvesting intensity approaches the threshold, the slower the population recovers to a new equilibrium after the next increase of the harvesting intensity. Mathematically, this follows from the fact that, as the harvesting intensity approaches the threshold, the eigenvalue of the linearization at the equilibrium goes to zero.

We should emphasize that under the considered conditions of harvesting it is dangerous to approach the maximum value of the yield (sometimes unjustifiably referred to as the optimal yield): if the harvesting intensity exceeds the maximum yield even by a small amount, then the population is doomed to die out. Therefore, such harvesting can hardly be thought of as optimal.

Let us use the same graphical technique to examine the second harvesting strategy where a constant fraction of individuals is removed from the population (see Fig. 2.3.1*b*). It can be seen that, depending on the value of α, there are two possibilities for the relative positions of the graphs of $F(x)$ and $S(x)$, and, correspondingly, for the dynamics. The behavior of a population at small harvesting intensity (see Eq. (2.3.3)) can be described by the logistic equation. For every value of α, the values for the parameters of the logistic equation are given by the expressions $\alpha' = a - \alpha$ and $K' = K(1 - \alpha/a)$. As the harvesting intensity grows, the equilibrium population size decreases monotonically, and for $\alpha > a$ the population is doomed to die out, whatever its initial size may be. The yield depends non-monotonically on the harvesting intensity.

This discussion leads to an important conclusion: although the maximum stationary yield of a population is independent of the harvesting strategies, the different strategies are not equal. Obviously the second strategy should be preferred where a definite fraction of the population is removed at a fixed harvesting intensity. In this situation, exceeding the harvesting intensity that corresponds to the maximum yield decreases the yield itself. This way one gets a warning that the optimal harvesting intensity has been exceeded. On the other hand, with the first harvesting strategy where the harvesting intensity is just a fixed number of individuals that are removed from the population per time unit, increasing the harvesting intensity above the maximum yield leads to the extinction of the population without a warning.

Considering other harvesting strategies with a more complicated relationships between removed individuals and the population size seems to be unnecessary since they are difficult to implement in practice.

Chapter 3

Predator-Prey Interactions

Three main types of interactions between species are recognized in ecology and are indicated by ++, −− and +− (Odum, 1971). A plus stands for a positive or favorable effect of one species upon another, and a minus for an unfavorable effect. The corresponding types of interspecifies interactions are known as

[++] protocooperation, mutualism or symbiosis;
[−−] mutual competitive suppression, or competition for a common resource;
[+−] predator-prey or parasite-host interactions.

Besides these main types, there are also interactions in which one species has either a positive or a negative effect on the other, but is completely unaffected by the other species (±0 type interaction).

In this notation, the + and − signs have a definite mathematical meaning that goes beyond their conventional metaphorical meanings. In particular, suppose that the dynamics of two interacting populations are described by a set of differential equations,

$$\dot{x} = xf(x,y), \qquad \dot{y} = yg(x,y)\,, \tag{3.0.1}$$

and the derivatives $\partial f/\partial y$ and $\partial g/\partial x$ have constant signs over the entire range of variables. Then the combination of the signs of these derivatives determines the nature of the the interactions between these two populations according to the classification outlined above. From this viewpoint, interactions of type ±0 are regarded as an exceptional, degenerate situation, and they will not be dealt with here.

We first consider trophic, or predator-prey, interactions, since they play the most important role in the functioning of ecosystems.

3.1. Volterra's Model and its Modifications

The first model to describe the size (density) dynamics of two populations interacting as a predator–prey system was suggested independently by A. Lotka (1925) and V. Volterra (1931):

$$\dot{x} = ax - bxy, \qquad \dot{y} = -cy + dxy\,, \tag{3.1.1}$$

In this equation x and y denote prey and predator densities, respectively, a is the reproduction rate of the prey population in the absence of the predator, b is the per capita rate of the consumption of prey by the predator population, c is the natural mortality rate of the predator, and d/b is the fraction of prey biomass that is converted into predator biomass.

The following idealizations of the inter- and intraspecies interactions in a predator–prey system form the basis of the model:

1. in the absence of the predator, the prey population grows exponentially according to Malthus' law;
2. if there is no prey, the predator population dies out exponentially;
3. the total amount of prey eaten by the predator per time unit depends linearly on the population densities of both predator and prey;
4. the portion of prey biomass that is converted into predator biomass is constant;
5. no other factors affect the dynamics.

In the initial notation (3.1.1), the system depends on four parameters, a, b, c and d. However, setting $x = \frac{a}{d}\,u$, $y = \frac{a}{b}\,v$, $t = \frac{\tau}{a}$ we can rewrite this as

$$\begin{aligned} \dot{u} &= u - uv\,, \\ \dot{v} &= -\gamma v + uv \end{aligned} \tag{3.1.2}$$

with the single parameter $\gamma = c/a$.

The analysis of this system in the first quadrant of phase space is well-known. There are two equilibria, a saddle at the origin and a center at the point $u = \gamma, v = 1$. The system is conservative, and all of its trajectories form closed orbits for any $\gamma > 0$.

In system (3.1.2), and in all of its modifications to follow, only the positive and finite values of variables admit biological interpretation. Nonetheless, it is often useful to understand how the system behaves when the variables are negative or tend to infinity. Figure 3.1.1 shows the complete phase portrait of system (3.1.2) on the Poincaré sphere.

The distinguishing feature of the Lotka–Volterra system is that it allowed drawing important conclusions regarding the qualitative behavior of a system using purely mathematical methods and very simple assumptions about the system. In particular, it led to conclusions about the dynamics of population densities that would not have been drawn without the construction and analysis of a proper mathematical model. Because of this success, it has become a classical system that forms a standard basis for many subsequent models in mathematical ecology.

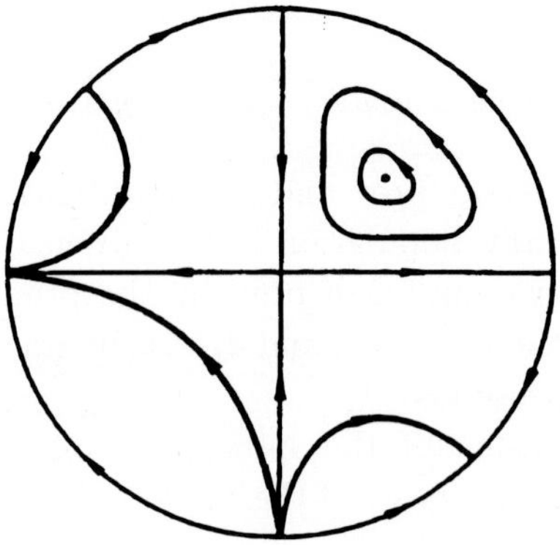

Fig. 3.1.1. Schematic diagram of phase portrait of system (3.1.2) on the Poincaré sphere for $\gamma > 0$.

At the same time, the model suffers from two fundamental, interrelated drawbacks which, in retrospect, could have been presented as sources for improvement and further development. Viewed mathematically, system (3.1.1) is structurally unstable according to the definition given by Andronov and Pontryagin (Andronov *et al.*, 1966). Furthermore, it is conservative so that any additional factors put into the model qualitatively changes its behavior. From a biological perspective, the drawback of this model is that it does not demonstrate the characteristic properties of any pair of populations that interact as a predator–prey system, for example predator saturation, limited predator and prey resources even if the prey is abundant.

The work of Volterra and Lotka stimulated research on the dynamics of interacting populations that developed in two directions. One was the aforementioned work of Kolmogorov (1936, 1972), whose principal idea was that functions of the model system should only be described on the basis of their qualitative properties, such as positivity, monotonicity, and the "more than" and "less than" relations $\langle$and$\rangle$.

In the second line of work, however, researchers successively examined specific modifications of system (3.1.1) that could be obtained by including a diversity of additional factors and relationships, described by explicitly assigned functions.

Several other articles appeared in which some of the functions in the model were given by an explicit formula, while the others were stated only in terms of some general assumptions of the above type. The well-known model of Rosenzweig–MacArthur (1963) is an example of this type of "hybrid" approach.

Research in each of these directions has advantages and shortcomings, and the hybrid approach is not immune either. For models of the first type, it is possible to formulate and prove some general assertions related chiefly to the number of equilibria and their local stability, but one cannot obtain a general understanding of the dynamics. In particular, one cannot determine the configuration of basins of attraction for individual equilibria and stable cycles. Models of the second type allow a more complete study, but the results obtained are less general.

A promising approach is to combine both directions in the following stages.

1. determine the main biological factors and relationships that should be taken into account by the model of a predator–prey system, and choose appropriate mathematical functions.
2. analyse the dynamic effects that result from incorporating these factors and relationships one at a time.
3. build and study a set of models for the predator–prey system that contain various combinations of the main biological factors affecting the dynamics.
4. finally, reveal features that are common to different models, and formulate general assertions about the nature of the dynamics in such a system.

Mathematically, the above program consists of the following. Consider the system

$$\begin{aligned} \dot{x} &= A(x) - B(x,y)\,, \\ \dot{y} &= -C(y) + D(x,y)\,, \end{aligned} \tag{3.1.3}$$

which is a generalization of system (3.1.1) in which all the terms are replaced by functions of the corresponding population densities. First, we list the important biological factors and relationships that have not been considered in the basic model (3.1.1). Then we analyse explicit formulas for the functions appearing in (3.1.3) which can be used to describe these relationships. We intend to study various modifications of system (3.1.3) by taking into account different combinations of these additional factors.

3.2. Elementary Factors of Interactions

This section examines some of the main biological factors in a predator–prey system with fundamental impacts on inter- and intrapopulation interactions.

The dynamics of isolated populations were studied in detail in Chapter 1. Some of the relationships discussed there appear later in the form of the function $A(x)$ in the analysis of system (3.1.3). For now, note that this allows for the incorporation of two new factors which were not a part of system (3.1.1). They involve the nonlinear character of the reproduction of the prey population at low densities, and intraspecies competition within the prey population that is induced by limited resources.

3.2.1. *Predation*

The function $B(x,y)$ describes the rate at which prey is consumed, that is, the predation rate, and it depends on the population densities of both prey and predator. It is an obvious and experimentally confirmed fact that $B(x,y)$ is a monotonically non-decreasing function of either argument. Therefore, it seems natural to first consider this function for one variable at a time while holding the other fixed, and then construct a general function of both variables.

A. *Trophic predation function.* Let the predator population density be fixed, for example equal to one, and consider the dependence of the predation rate on the density of prey. We denote this function by $B(x, \cdot)$. In ecology, this dependence is known as the trophic predation function or the functional reaction of the predator to the prey population density (Holling, 1965). In model (3.1.1) this is simply the linear function $B(x, \cdot) = bx$. This points to the absence of predator saturation in (3.1.1), since, for instance, doubling the amount of food which is available to the predator doubles its consumption. It is clear that such a dependence is only valid at relatively small prey densities.

A detailed theoretical and experimental investigation of the form of this trophic function was carried out by Ivlev (1955) who proposed the following dependence:

$$\rho = R(1 - e^{-\xi p}) . \tag{3.2.1}$$

Here, ρ is the predator's consumption, that is, the biomass of prey consumed by the predator per time unit, R is the limit consumption, that is, the predator's consumption when prey is abundant, p is the prey population density or the measure for the quantity of food available to the predator, and ξ is a constant which has a dimension of the inverse of the population density. In the notation used above, equation (3.2.1) can be rewritten as

$$B(x, \cdot) = B_{\max}(1 - e^{-\xi x}) . \tag{3.2.2}$$

The same form for the dependence was put on a theoretical foundation by N. Rashevsky who reasoned from considerations concerning feeding mechanisms. In the field of microbiology, the dependence of the rate of comsumption of a substrate by a microorganism on the concentration of substrate is described by the following formula suggested by Monod (1942):

$$B(x, \cdot) = \frac{bx}{1 + \alpha x} . \tag{3.2.3}$$

This formula can also be used to describe trophic functions for the predator population. Here, $b/\alpha = B_{\max}$ is the maximum consumption of the predator, α is a constant whose dimension is the inverse of the population density and that plays a similar role as the constant ξ in (3.2.1): $1/\alpha$ is the prey population density at which the predator's consumption is half the maximum value (Bazykin, 1974).

Holling (1965) noticed that the trophic function might be qualitatively different from the linear one, not only at large densities of prey (because of predator saturation), but also at small prey densities. In this case, the curve $B(x, \cdot)$ is tangent to the abscissa at the origin. An explicit formula of such a function is for example

$$B(x, \cdot) = \frac{bx^2}{1 + \alpha x^2} , \tag{3.2.4}$$

or in a more general form

$$B(x, \cdot) = \frac{bx^2}{1 + \alpha_1 x + \alpha_2 x^2} . \tag{3.2.5}$$

There can be two reasons for such a behavior of the trophic function at small prey densities: either the predator has an alternative source of food or the prey has a number of shelters inaccessible to the predator (Holling, 1965).

Thus, following Holling we distinguish three types of trophic functions which can be presented qualitatively as shown in Fig. 3.2.1. Here, it is correct to interpret the third type of trophic function as a result of the joint consideration of two elementary factors: nonlinear dependence of the predation rate on the prey population density, and predator saturation at large prey population densities.

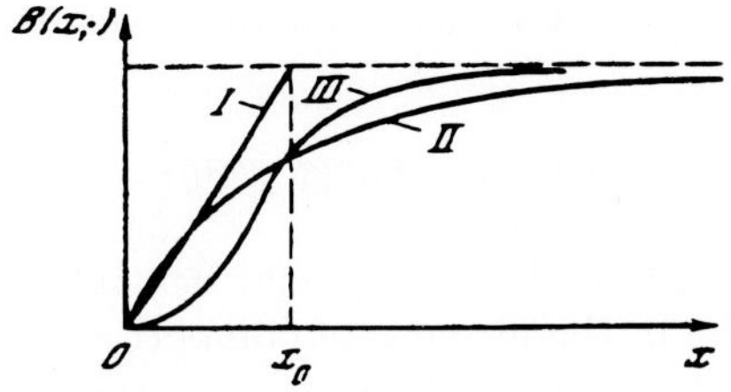

Fig. 3.2.1. Three types of trophic functions after Holling (1965).

In Holling's classification, the first type of reaction of the predator to the prey population density is a piece-wise linear function with a slope in a range $x_0 > x > 0$, and horizontal for values of x exceeding x_0 (see Fig. 3.2.1). In this book we only consider smooth functions and, for the sake of convenience, assume that functions of this first type are simply linear.

B. *Predator competition for prey.* Let us now proceed to study the dependence of the predation rate on the predator population density for some fixed density of prey. We denote this function by $B(\cdot, y)$. In the initial model (3.1.1), this function, as well as the predator trophic function, was linear, so that $B(\cdot, y) = by$.

Predator trophic functions and their application to model predator–prey systems have been extensively examined in the literature, but the relationship between the predation rate and the predator population density has been given less consideration. In most cases, it is simply written as

$$B(x, y) = yB(x)\,. \tag{3.2.6}$$

It is obvious that the competition among predators for prey is excluded from consideration here. In fact, in this notation the rate of prey consumption per unit of predator density, or per predator, is independent of the density of the predator population. In other words, it is assumed that individual predators do not interact with each other, and in particular, do not compete. A similar approach to describe food dynamics in microbiology has proved to be completely accurate. In that context, the predator is a population of microorganisms present in a nutrient solution and the prey is the nutrient. This approach however, can hardly be applied to ecologal problems as the situation there is generally different. Therefore, it may be reasonable to regard Eq. (3.2.5) as the asymptotic case describing a very low intensity of competition for prey, that is, corresponding to the vanishing of the predator.

We may also consider the opposite extreme of fierce competition, in which the rate of consumption per predator is proportional to the number of prey available to this predator, rather than to the total size of the prey population. Such a situation was first considered by Leslie (1948). It is obvious that in this case the total rate of prey consumption is independent of the predator population density.

We may assume that competition among predators for prey is negligible when the predator population density tends to zero, and fierce when the predator population density grows without bound. For intermediate population densities, we assume the following dependence:

$$B(\cdot, y) = by/(1 + \beta y)\,, \tag{3.2.7}$$

where $1/\beta$ is the predator population density at which the predator's consumption is half of what it would be in the absence of competition for prey.

Naturally, there are other ways to describe predator competition for prey. However, our approach is simple, convenient and does not contradict the experimental data. The representation of (3.2.6) has the following schematic interpretation: at small population densities, the individual predators do not hinder each other and catch the prey independently; at large densities, they remove as many prey as possible from the prey population at given prey density, and the further growth of the predator population does not increase the total amount of prey consumed by all predators.

C. *General form of the predation function.* We have described the possible forms of the function $B(x, y)$ with either argument fixed. What should this function really look like? The author is unaware of specific experimental work on the interrelation of predator saturation and predator competition for prey, although the facts cited by Ivlev (1955) may provide indirect evidence that these effects are independent over a sufficiently wide range of predator and prey population densities. In this case, the function $B(x, y)$ takes the form

$$B(x, y) = B(x, \cdot) \cdot B(\cdot, y)\,. \tag{3.2.8}$$

We sum up the above discussion of which function should be used to describe the dependence of predation rate on predator and prey population densities: in first approximation, corresponding to the classical Volterra scheme (equations (3.1.1)), it is natural to use a function which is linear in both arguments, that is, bilinear. In order to make this function more specific, it is useful to take also the following factors into consideration:

1. predator saturation, which corresponds to Holling's type II trophic function.
2. nonlinear (quadratic) dependence of the predation rate on the prey population density at low levels. (Assuming nonlinearity of predator saturation, this corresponds to Holling's type III trophic function.)

3. predator competition for prey, that is, a decrease in the per predator rate of consumption with an increase in the predator population density for fixed prey density.

The first two of these factors refer to the dependence of the rate of predation on the density of prey, while the third factor deals with the dependence of the predation rate on the predator population density.

Hence, depending on which factors we take into account, the function $B(x, y)$ can be described by means of one of the functions presented in Table 1.

Table 1

	Type of Trophic Function		
Competition	I	II	II
No	bxy	$\dfrac{bxy}{1+\alpha x}$	$\dfrac{bx^2y}{1+\alpha x^2}$
Yes	$\dfrac{bxy}{1+\beta y}$	$\dfrac{bxy}{(1+\alpha x)(1+\beta y)}$	$\dfrac{bx^2y}{(1+\alpha x^2)(1+\beta y)}$

3.2.2. *Reproduction and Mortality of the Predator*

Ivlev (1955) pointed out that the dependence of the population growth rate on feeding habits is very complex. Recall that the initial model (3.1.1) assumes the existence of a constant coefficient of conversion of prey biomass into predator biomass. We retain this, as a starting assumption, and only introduce one additional factor. In analogy to the consideration of the growth dynamics of an isolated population, we allow for the fact that at small predator densities the reproduction rate may be limited by the lack of potential breeding partners, rather than by the shortage of food. Therefore, this rate is taken to be proportional, not to the population density, but to the square of this value (see Sec. 2.1). In this case, the fertility function $D(x, y)$ takes the form

$$D(x,y) = \frac{dy}{N_y + y} \cdot B(x,y)\,, \tag{3.2.9}$$

where N_y is the predator population density at which the reproduction rate is half of the maximum rate achieved when prey is abundant.

In model (3.1.1), predator mortality is assumed to be constant, that is, independent of the predator density. This assumption may be suplemented to allow for competition among predators for resources other than prey. When describing the form of the function $B(x, y)$, we have only analysed the character of predator competition for prey. In reality, however, a predator population may also be limited by shortage of other limited resources such as the size of the habitat suitable for the predator to live and reproduce in. In this case, as in the description of single-population dynamics, it is natural to allow for predator competition by introducing a negative quadratic term into the equation for the rate of change in predator density. Hence, the function $C(y)$ takes the form

$$C(y) = -cy - hy^2 , \tag{3.2.10}$$

where h is the coefficient of competition for resources other than prey.

3.2.3. *List of Elementary Factors*

Using the classical system (3.1.1) as a first approximation to the dynamics of two populations interacting as a predator–prey system, we have consecutively considered the following additional factors:

1. nonlinear (quadratic) dependence of the reproduction rate of the prey population on the prey population density when this density is low,
2. competition among prey,
3. mortality of prey (in the case when nonlinear reproduction at small population densities is taken into account),
4. predator saturation,
5. nonlinear (quadratic) dependence between the predation rate and the prey population density at small prey densities,
6. predator competition for prey,
7. predator competition for resources other than prey, and
8. nonlinear (quadratic) dependence between the predator reproduction rate and the population density at small prey densities.

From a mathematical point of view, when we analyse the dynamic effects of introducing each separate factor into the model, we deal with a one-parameter perturbation of system (3.1.1). If a pair of factors is considered simultaneously, we deal with a two-parameter perturbation, and so on. Thus, in a formal sense, a study of the joint effect that all of the enumerated factors have on the dynamics of predator–prey system is in fact a complete qualitative investigation of an eight-parameter perturbation of system (3.1.1), that is, an analysis of a system which, after scaling, depends on nine parameters. However, dividing the nine-dimensional parameter space into domains that correspond to qualitatively different types of system behavior, is not only unrealistic but also meaningless. Inevitably, the results of such a study would be bulky and uninterpretable. Therefore, we intend to proceed by gradually complicating the study, and we begin with the analysis of one-factor modifications of system (3.1.1). After that, the results are used to classify the factors considered, in order to specify which of the two-factor modifications and which of the more complex modifications can provide the most comprehensive pictures of the dynamic behavior of predator–prey systems.

3.3. One-Factor Modifications of the Volterra Model

Many researchers have analysed one-factor modifications of system (3.1.1) analogous to those that interest us. In some sense, it is not worth bothering with this analysis here, because there always exists a unique nontrivial equilibrium, whose stability can be unambiguously determined from the relative positions of the nullclines (Rosenzweig, MacArthur, 1963). Therefore, we simply write down the respective systems of differential equations in their initial and scaled forms, and present their schematic phase portraits. Note that, since the initial system (3.1.2) only depends on one parameter, its one-factor modifications depends on two parameters.

3.3.1. *Nonlinear Reproduction, Competition and Mortality of Prey*

Incorporating nonlinear reproduction of the prey population leads to the system

$$\begin{aligned} \dot{x} &= \frac{ax^2}{N+x} - bxy\,, \\ \dot{y} &= -cy + dxy\,, \end{aligned} \tag{3.3.1}$$

where N denotes the prey density at which the reproduction rate is half of its possible maximum value. Setting $t = \tau/a$, $x = Nu$ and $y = (a/b)\ v$ transforms system (3.3.1) into

$$\begin{aligned} \dot{u} &= \frac{u^2}{1+u} - uv\,, \\ \dot{v} &= -\gamma v + \kappa uv\,, \end{aligned} \tag{3.3.2}$$

with parameters $\gamma = c/a$ and $\kappa = dN/a$.

The phase portrait of system (3.3.2) exhibits exactly one unstable equilibrium A in the first quadrant for all parameter values (Fig. 3.3.1). All trajectories spiral away to infinity. (The limit can be thought of as an infinitely far limit cycle)

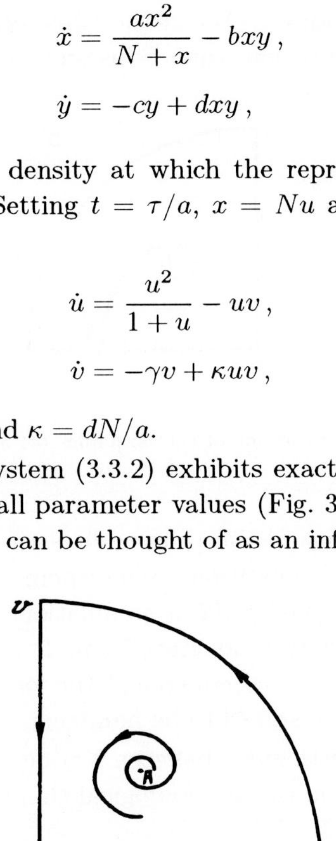

Fig. 3.3.1. Phase portrait of system (3.3.2).

Taking prey competition into account leads to the system

$$\begin{aligned} \dot{x} &= ax\frac{K-x}{K} - bxy\,, \\ \dot{y} &= -cy + dxy\,, \end{aligned} \tag{3.3.3}$$

which, by setting $t = \tau/a$, $x = Ku$ and $y = (a/b)v$, gives

$$\begin{aligned} \dot{u} &= u(1-u) - uv\,, \\ \dot{v} &= -\gamma u + \kappa uv\,. \end{aligned} \tag{3.3.4}$$

Two relative positions of the nullclines and, correspondingly, two phase portraits of the system are possible (Fig. 3.3.2). When $\gamma/\kappa < 1$, there exists a stable equilibrium A at $\{u = \gamma/\kappa, v = 1 - \gamma/\kappa\}$ inside the first quadrant and a saddle B at $\{u = 1, v = 0\}$ on the abscissa (Fig. 3.3.2*b*). As γ/κ increases the equilibria A and B approach each other, and merge at $\gamma/\kappa = 1$, to form a stable saddle-node with nodal sector in the first quadrant. When the value of γ/κ is increased further, A disappears into the negative region while B becomes a stable node (Fig. 3.3.2*c*).

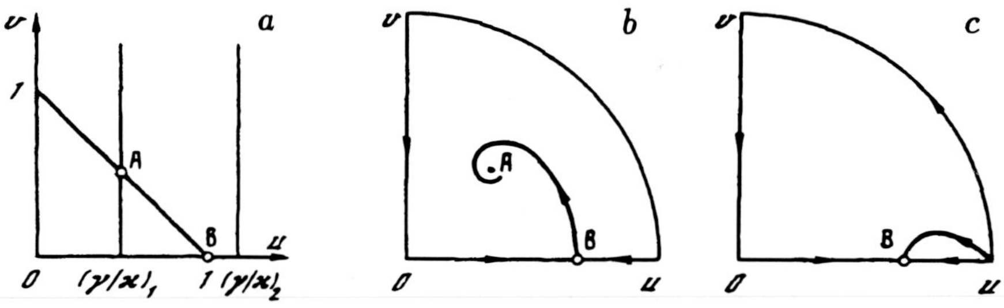

Fig. 3.3.2. Two possible relative positions of the nullclines (a) and the phase portraits (b and c) of system (3.3.4).

This suggests the inability of the prey to feed the predator population when there is strong intraspecies competition among them, that is, when they are short of resources. In this case, the predator population is doomed to die out from whatever initial state the system may start from. It follows from Sec. 2.1 that the incorporation of prey mortality does not change the form of system (3.1.1) as long as the prey reproduction is not assumed to be nonlinear at small population densities.

Note that we shall not distinguish between a node and a focus, unless otherwise stated, and only pay attention to the topological character of an equilibrium.

3.3.2. *Predator Saturation (Type II Trophic Function)*

If we allow for predator saturation in the model, we obtain the system

$$\begin{aligned} \dot{x} &= ax - \frac{bxy}{1 + Ax}\,, \\ \dot{y} &= -cy + \frac{dxy}{1 + Ax}\,. \end{aligned} \tag{3.3.5}$$

Setting $t = \tau/a$, $x = (a/d)u$ and $y = (a/b)v$ transforms this system into

$$\dot{u} = u - \frac{uv}{1 + \alpha u}, \qquad \dot{v} = -\gamma v + \frac{uv}{1 + \alpha u}. \tag{3.3.6}$$

The nullcline equations $\dot{u} = 0, v = 1 + \alpha u$ and $\dot{v} = 0, u = \gamma/(1 - \alpha\gamma)$ show that an equilibrium only exists for $\alpha\gamma < 1$. There is a very natural interpretation of this formal result. In particular, the maximum predator growth rate is $1/\alpha$ when prey is abundant. Therefore, if $\alpha\gamma > 1$ the derivative $\dot{y}$ is less than 0 over the entire range of the variables, there does not exist a nontrivial equilibrium, and the predator population is doomed to die out. For $\alpha\gamma < 1$, the equilibrium exists inside the first quadrant, but it is always unstable. The phase portraits of the system for different parameter values are shown in Fig. 3.3.3.

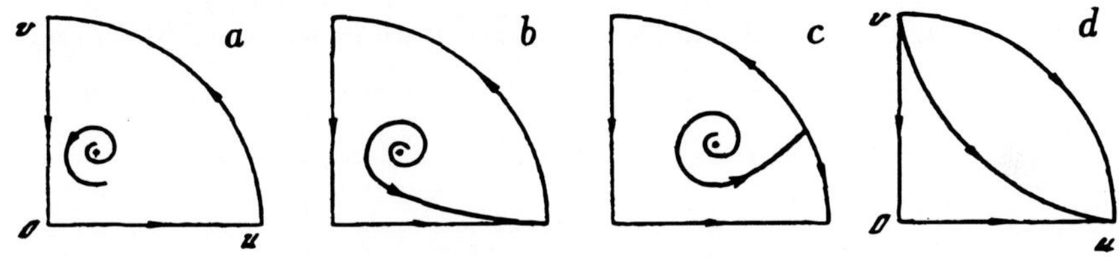

Fig. 3.3.3. Possible phase portraits of system (3.3.6) for $\alpha\gamma < 1$ (*a*–*c*) and for $\alpha\gamma > 1$ (*d*).

We now focus on a peculiarity of the system's behavior near infinity. When $\alpha = 0$, the system is conservative and its trajectories are cloesd curves. For $1/(1 + \gamma) > \alpha > 0$, the equilibrium becomes unstable and the infinitely far limit cycle turns out to be attracting (see Fig. 3.3.3*a*). This indicates that the prey and predator population densities vary with an amplitude that increases without bound. When the predator population reaches its peak density and gains again a control over the prey population it drives the latter close to extinction, which becomes more pronounced with each round.

However, as α grows further, a change in the behavior of the system takes place at infinity. When $\alpha = 1/(1 + \gamma)$ the point corresponding to the end of the abscissa becomes a saddle-node (Fig. 3.3.3*b*), and for $\alpha > 1/(1 + \gamma)$ it becomes a globally attracting node at infinity and a saddle emerges "above" it on the boundary of the Poincaré disk (Fig. 3.3.3*c*). This may be interpreted as follows. An increase of this parameter corresponds to a drop in the predator's biotic potential, that is, the maximum possible reproduction rate of the predator population. For $\alpha > 1/(1+\gamma)$ and larger prey population densities the predator is unable to overtake the prey and drive it back to lower densities. As a result, the prey population density grows monotonically without bound, and the predator population increases as well. This phenomenon in which the prey escapes the predator has been described for a completely different model (Takahashi, 1964).

3.3.3. *Nonlinear Predation at Small Prey Population Density*

The predator–prey interaction described by the type III trophic function is characterized by a nonlinear increase in the absolute predation rate with the growth of prey population density, and by the predator saturation effect we have just considered. In particular, we have

$$\dot{x} = ax - \frac{bx^2y}{1+Px}, \qquad \dot{y} = cy + \frac{dx^2y}{1+Px}, \tag{3.3.7}$$

where P is a quantity that is inversely proportional to the prey density. This system describes the quadratic character of the predation rate when the prey population is small, and asymptotically approaches the initial system (3.3.1) at larger prey population densities.

Setting $t = \tau/a$, $x = \sqrt{a/d}u$ and $y = (\sqrt{ad}/b)v$ we can rewrite (3.3.7) as

$$\dot{u} = u - \frac{u^2v}{1+\alpha u}, \qquad \dot{y} = -\gamma v + \frac{u^2v}{1+\alpha u}, \tag{3.3.8}$$

with the new parameters $\gamma = c/a$ and $\alpha = P\sqrt{a/d}$. A nontrivial equilibrium for this system always exists and is stable for all values of the parameters (Fig. 3.3.4).

3.3.4. *Predator Competition for Prey and Other Resources*

Competition among predators for prey can be represented by the system

$$\dot{x} = ax - \frac{bxy}{1+By}, \qquad \dot{y} = -cy + \frac{dxy}{1+By}. \tag{3.3.9}$$

Setting $t = \tau/a$, $x = (a/d)u$, and $y = (a/b)v$ this becomes

$$\dot{u} = u - \frac{uv}{1+\beta v}, \qquad \dot{v} = -\gamma v + \frac{uv}{1+\beta v}, \tag{3.3.10}$$

where $\beta = aB/b$.

For $\beta < 1$, a nontrivial equilibrium exists and is stable, and the phase portrait is analogous to that in Fig. 3.3.4. As β increases, the equilibrium sizes of predator and prey populations grow. However, when $\beta > 1$, the growth rate of the prey population exceeds the rate of predation at arbitrarily large predator population densities. The prey population grows without bound, and the nontrivial equilibrium is absent; the phase portrait is like that in Fig. 3.3.3*d*.

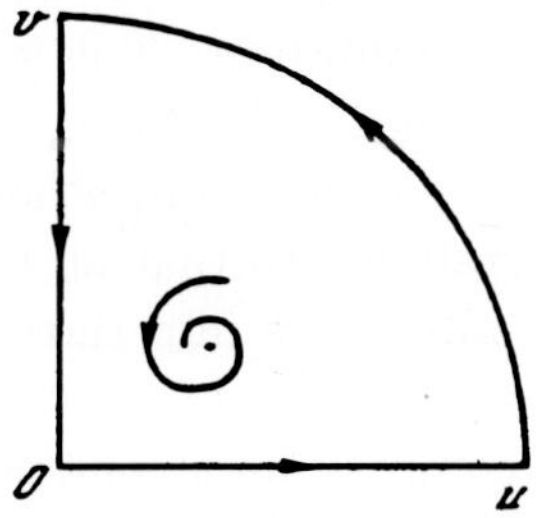

Fig. 3.3.4. Phase portrait of systems (3.3.8) and (3.3.10) for $\beta < 1$.

Allowing for predator competition for resources other than prey leads to the following system:

$$\begin{aligned} \dot{x} &= ax - bxy\,, \\ \dot{y} &= -cy + dxy - hy^2\,. \end{aligned} \tag{3.3.11}$$

Note that if we multiply both equations of (3.3.9) by $(1 + By)$, that is, make an appropriate scaling of time, we obtain the system

$$\begin{aligned} \dot{x} &= ax - (b - aB)xy\,, \\ \dot{y} &= -cy + dxy - Bcy^2\,, \end{aligned} \tag{3.3.12}$$

which agrees with (3.3.11) after a parameter transformation. Therefore, the phase portrait of (3.3.11) is identical to that of Fig. 3.3.4.

3.3.5. *Nonlinear Reproduction of the Predator at Small Population Densities*

If we take into account the nonlinearity in the predator population's reproduction rate when the density is small, the model becomes

$$\begin{aligned} \dot{x} &= ax - bxy\,, \\ \dot{y} &= -cy + \frac{y}{N + y}\,dxy\,. \end{aligned} \tag{3.3.13}$$

Setting $t = \tau/a$, $x = (aN/d)u$ and $y = (a/b)v$ yields

$$\begin{aligned} \dot{u} &= u - uv\,, \\ \dot{v} &= -\gamma v + \frac{uv^2}{1 + \nu v}\,, \end{aligned} \tag{3.3.14}$$

with $\gamma = c/a$ and $\nu = a/bN$. The equations for the nullclines are $\dot{u} = 0$, $v = 1$ and $\dot{v} = 0$, $u = \gamma(1 + \nu v)/v$. A nontrivial equilibrium always exists and is unstable. All trajectories escape to infinity, and the phase portrait is like that of Fig. 3.3.1.

3.3.6. *Classification of Elementary Factors*

The above analysis leads to the following conclusion. A modification of the Volterra system (3.1.1) that accounts for one of the factors mentioned in the preceding section changes the nature of the only equilibrium. This equilibrium corresponds to the neutrally stable, that is, not asymptotically stable, coexistence of the predator and prey populations. In particular, the equilibrium either attains stability and becomes globally attracting, or it loses stability, so that all trajectories go to infinity. This makes it possible to subdivide all factors into those that stabilize and those that destabilize the equilibrium.

The stabilizing factors are: competition among prey, competition among predators for prey or other resources, and nonlinearity in the trophic function when the prey population density is small. The destabilizing factors are: predator saturation and nonlinear reproduction of the predator and prey populations at small densities.

It is not difficult to prove that the simultaneous consideration of two or more stabilizing factors, or of two or more destabilizing factors, does not lead to any new results. New equilibria do not appear, and the combination of stabilizing factors always yields stability of the unique equilibrium, whereas the combination of destabilizing factors leads to instability. Thus, it is necessary to study only combinations of stabilizing with destabilizing factors, which is what we do in the next section.

3.4. Two-Factor Modifications of the Volterra Model

The initial system (3.1.1), scaled to the form (3.1.2), depends on only one parameter γ and remains conservative for all values of this parameter. The incorporation of individual factors that affect the dynamics of predator and prey population densities has resulted in one-factor modifications of the initial system (3.1.2), that is, in systems that depend on two parameters. One might expect that the corresponding two-parameter plane would contain curves of bifurcations of higher codimension that divide this plane into regions in which the system exhibits different qualitative behavior. However, the previous section has revealed that the qualitative behavior of such a system is either qualitatively unchanged for all parameters or changes qualitatively only when the "perturbing" parameter is changed, independently of the parameter γ in the initial system (3.1.2).

The simultaneous consideration of a pair of competing factors, a stabilizing and a destabilizing one, leads to two-factor modifications of system (3.1.2), or to systems that depend on three parameters. In general, finding the division of a three-dimensional parameter space into regions of qualitatively different dynamic behaviors of the system is not a simple task. Therefore, to get a three-parameter diagram, we assume that the parameters are not equivalent, and we study the structure of the two-dimensional parameter space of the "perturbing" parameters by fixing the value of the parameter γ inherited from the initial system (3.1.2). More precisely, we intend to construct the bifurcation diagram in the form of a one-parameter family of two-parameter cross sections given by $\gamma =$ constant.

Our further analysis of concrete systems, to our best abilities, proceeds as follows.

1. we write down and scale the initial form of the system,
2. analyse the system mathematically,
3. describe its bifurcation diagram and phase portraits,
4. describe the dynamic regimes that are realized in the system, as well as the reaction of the system to perturbations of the phase variables,
5. characterize the evolution of dynamic regimes under changes of the parameters, and, in particular, when parameters cross bifurcation surfaces, and
6. interpret the results from the ecological point of view.

Now, let us proceed with a detailed study of systems that describe the effects that each pair of competing factors has on the Volterra system (3.1.2).

3.4.1. *Prey Competition and Predator Saturation*

If we include both, predator saturation and competition among the prey, the model becomes

$$\begin{aligned} \dot{x} &= ax\frac{K-x}{K} - \frac{bxy}{1+Ax}, \\ \dot{y} &= -cy + \frac{dxy}{1+Ax}, \end{aligned} \tag{3.4.1}$$

which is a particular case of the system suggested by Rosenzweig and MacArthur (1963). The existence of a stable limit cycle in this model in some parameter region was shown almost simultaneously by May (1972), Shimazu *et al.* (1972) and Gilpin (1972). Later, the system was further studied by Kasarinoff and Deiesch (1978). For the sake of consistency, we shall study it in terms of the parameters used throughout the present book.

Setting $t = \tau/a$, $x = (c/d)u$ and $y = (a/b)v$, system (3.4.1) becomes

$$\begin{aligned} \dot{u} &= u - \frac{uv}{1+\alpha u} - \epsilon u^2, \\ \dot{v} &= -\gamma v\left(1 - \frac{u}{1+\alpha u}\right), \end{aligned} \tag{3.4.2}$$

with parameters $\alpha = Ac/d$, $\epsilon = c/Kd$, and $\gamma = c/a$. The equations for the nullclines are $\dot{u} = 0$, $v = (1+\alpha u)(1-\epsilon u)$ and $\dot{v} = 0$, $u = 1/(1-\alpha)$. From Fig. 3.4.1*a*, one can easily determine the region of existence of a nontrivial equilibrium: it is obvious that the nullclines intersect in the first quadrant if $\alpha + \epsilon < 1$. The region of an unstable equilibrium is the region in which the vertical nullcline $\dot{v} = 0$ is located on the left of the maximum of the nullcline $\dot{u} = 0$, that is, when $\epsilon < \alpha(1-\alpha)/(1+\alpha)$. The corresponding curves in the (α, ϵ)-plane form the bifurcation diagram of the system (Fig. 3.4.1*b*). The first Lyapunov quantity (or the third focus quantity) (Bautin, Leontovich, 1976) is negative along the entire Andronov-Hopf curve

$\epsilon = \alpha(1-\alpha)/(1+\alpha)$. This means that, if the parameters pass from region *2* to region *3*, the equilibrium loses its stability in an Andronov-Hopf bifurcation, and a stable limit cycle appears around it. Note that the qualitative behavior of system (3.4.2) is independent of the parameter γ. The complete set of phase portraits is shown in Fig. 3.4.1*c*.

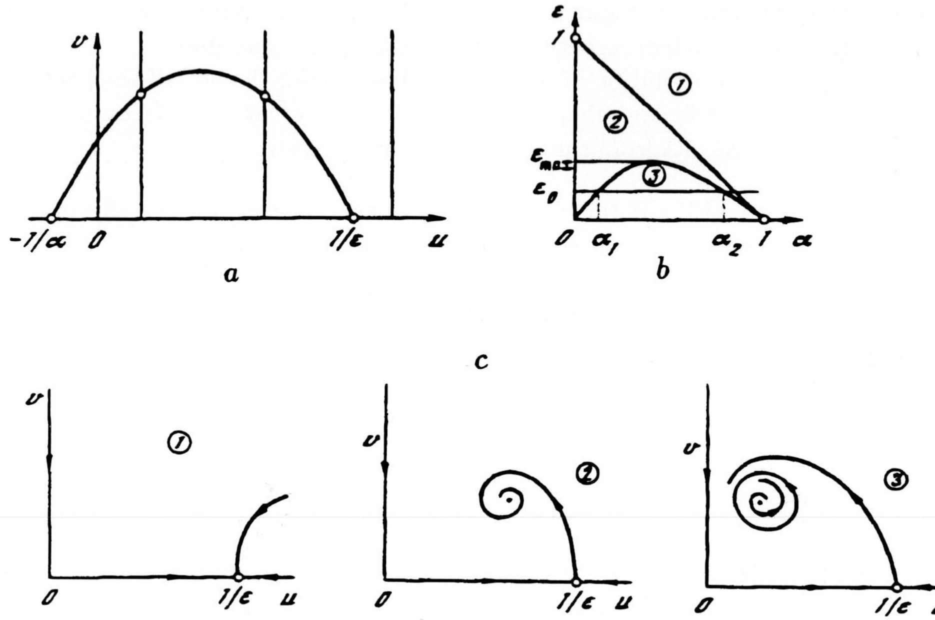

Fig. 3.4.1. Three possible relative positions of nullclines (*a*), the bifurcation diagram (*b*), and phase portraits (*c*) of system (3.4.2). (Throughout the text, the labeling of regions in parameter space matches the labeling of phase portraits.)

Summing up, we can say that allowing for the joint action of the stabilizing pressure of prey competition and the destabilizing force of predator saturation reveals two regions in the parameter plane that characterize the intensity of these factors: a region with a stable equilibrium, above the Andronov-Hopf curve $\epsilon = \alpha(1-\alpha)/(1+\alpha)$, where the first factor dominates, and a region with an unstable equilibrium, below the Andronov-Hopf curve, where the predator and prey populations can only coexist in an oscillatory fashion. Thus, a gradual weakening of a stabilizing factor, in this case a decrease in ϵ, may cause the equilibrium to lose its stablility in a gradual excitation of oscillations.

It should be noted that, although prey competition is always considered as a stabilizing factor, predator saturation is not an unconditionally destabilizing one, if both factors are analysed simultaneously. Indeed, consider what happens when we vary the parameter α for a fixed value of the parameter $\epsilon < \epsilon_{\max} \approx 0.17$ (Fig. 3.4.1*b*). If $\alpha < \alpha_1$, then the equilibrium is stable. As α grows, it passes the threshold α_1, and a small stable limit cycle appears in the phase portrait that corresponds to gradual excitation of oscillations. As α grows further, the size of this limit cycle increases to some maximum value and then decrease until α reaches a threshold α_2, when it shrinks to a point, so that the equilibrium is stable again. Thus, an increase in the parameter α, describing the rate of predator saturation, may result in both the loss and the restoration of the stability of the equilibrium.

3.4.2. *Prey Competition and Nonlinear Reproduction of Prey at Small Population Densities*

Considering simultaneously competition among prey and nonlinearity in the prey population's reproduction rate for small population sizes leads to the system

$$\begin{aligned} \dot{x} &= \frac{ax^2}{N+x}\,\frac{K-x}{K} - bxy\,, \\ \dot{y} &= -cy + dxy \end{aligned} \tag{3.4.3}$$

(see Bazykin, Khibnik, 1981*a*,*b*). Setting $t = \tau/a$, $x = Ku$ and $y = (a/b)v$ this becomes

$$\begin{aligned} \dot{u} &= \frac{u^2(1-u)}{n+u} - uv\,, \\ \dot{v} &= -\gamma v(m-u)\,, \end{aligned} \tag{3.4.4}$$

where $n = N/K$, $m = c/dK$ and $\gamma = dK/a$.

Note that in this case, and in several later instances, it is convenient to make a change of variables so that the value $u = m$ at an equilibrium can be regarded as a parameter of the scaled system. In general, a system depending on several parameters can be scaled in many ways to obtain different dependences on parameters. Although all of these scaled forms are formally equivalent, some of them may be given preference for the sake of convenience and, more importantly, for getting more readily interpretable results. There are no clear recipes for scaling, so the choice of parameters depends on the researcher's intuition.

The nullclines of system (3.4.4) have the equations

$$\dot{u} = 0, \quad v = \frac{u(1-u)}{n+u}; \qquad \dot{v} = 0, \quad u = m\,.$$

Two examples of relative positions of nullclines are shown in Fig. 3.4.2*a*. The parameter (m, n)-plane of the system shows at least two bifurcation curves (3.4.2*b*). Along the curve $m = 1$, the saddle-node AB generates a stable node A as we cross into the interior of the first quadrant. The other curve, denoted by N, corresponds to the values of parameters at which the vertical nullcline for $\dot{v} = 0$ passes through the maximum of the nullcline for $\dot{u} = 0$. This bifurcation curve reflects the loss of stability of A in an Andronov-Hopf bifurcation. The equation for this Andronov-Hopf curve N is

$$n = m^2/(1-2m)\,.$$

Now consider what happens when we cross the Andronov-Hopf curve N in the (m,n)-plane. The focus A loses its stability in one of two possible ways: either a small stable limit cycle appears in the phase portrait, or a small unstable limit cycle shrinks to a point. Let us calculate the first Lyapunov quantity L_1 along the Andronov–Hopf curve N. For (3.4.4) L_1 is given by the expression

$$L_1 = -\frac{\pi}{4}\,\frac{(1-2m)(4m-1)}{m(1-m)^2},$$

which is positive on the Andronov-Hopf curve above the point $\mathcal{F}$ $\{m = 1/4, n = 1/8\}$, and negative below that point. Thus, when the Andronov-Hopf curve is intersected above this point, a small stable limit cycle appears in the phase portrait, and when that curve is intersected from right to left below this point, an unstable limit cycle shrinks to a point. This unambiguously implies the existence of one more bifurcation curve Q in the phase portrait which is located between the Andronov-Hopf curve and the vertical line $m = 1$. This curve corresponds to the appearance of a pair of limit cycles in a saddle-node of limit cycles, and is called the curve of saddle-nodes of limit cycles. One of the endpoints of Q is the point $\mathcal{F}$ on the Andronov-Hopf curve (Arnol'd, 1978) and the other is the origin of the (m,n)-plane.

In the parameter plane the curve of saddle-nodes of limit cycles is one of the boundaries of a region in which the phase portrait has two nested limit cycles, of which the outer one is stable. The Andronov-Hopf curve forms the other boundary for this region (Fig. 3.4.2*b*).

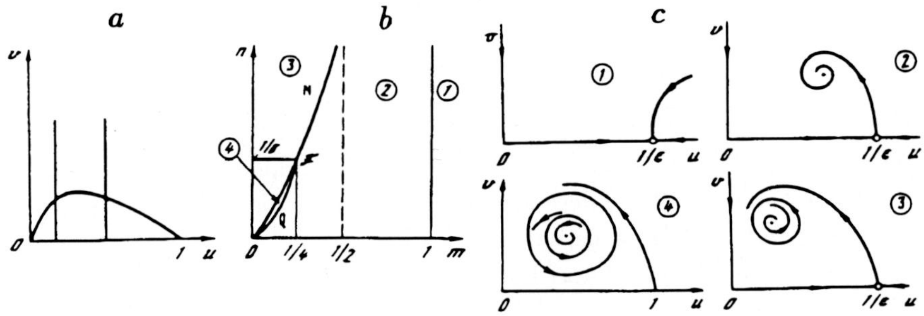

Fig. 3.4.2. Two possible locations of nullclines (*a*), the bifurcation diagram (*b*), and phase portraits (*c*) for system (3.4.4).

Two essential features should be mentioned. First, the qualitative behavior of system (3.4.4), just as in all of the preceding systems, does not depend on the parameter γ. Nevertheless, in contrast to the systems investigated so far, the dynamics is not completely determined by the relative positions of nullclines: when the vertical nullcline for $\dot{v} = 0$ is located on the right side of the maximum of nullcline for $\dot{u} = 0$, either a globally stable equilibrium A (Fig. 3.4.2*c*, region *2*), or a pair of stable cycles (Fig. 3.4.2*c*, region *4*) may exist in the phase portrait of the system.

Consider now what happens when the value of the parameter m is changed for a fixed $n < 1/8$. Suppose that initially $1 > m > 1/2$, so there exists one global attracting equilibrium A. As m decreases it intersects the curve of saddle-nodes of limit cycles. This does not affect the local stability of A and remains unnoticed to an observer who follows the changes of the system's equilibria. Only the basin of attraction changes, as it becomes bounded by the inner, unstable limit cycle. In this case, any sufficiently strong perturbation may drive the system out of the basin of attraction of the equilibrium and into an oscillatory mode.

As m decreases further, the basin of attraction of A becomes smaller until it shrinks to a point when m intersects the Andronov-Hopf curve. The equilibrium A becomes unstable, and the system trajectories spiral away from the equilibrium to the large limit cycle. This leads to abrupt excitation of oscillations.

Conversely, if m grows, the observer who keeps track of the oscillatory state of the system, does not notice the moment when this parameter intersects the Andronov-Hopf curve. In this case, the limit cycle does not disappear and retains its stability. As m continues to increase, oscillations suddenly become damped when m intersects the curve of saddle-nodes of limit cycles, and the system attains the equilibrium state.

For ecological applications, this hysteresis phenomenon, typical for abruptly generated oscillations, may be of great significance. If the parameter m is changed in one direction and then back again, *the same* parameter value may correspond to *different* dynamic regimes of the ecosystem. In other words, restoring the conditions, that is, the parameters, under which the system existed, does not guarantee that it returns to the previous state.

In physics, the term *abrupt excitation of oscillations* (also called *hard generation of oscillations*) is often used for two closely related, yet different things. In the phase-space sense, the term applies to the situation in which the phase portrait includes a stable equilibrium with a basin of attraction that is bounded by an unstable limit cycle, which in turn is surrounded by a stable limit cycle (Fig. 3.4.2*c*, region *4*). For small perturbations, damped oscillations restore the equilibrium, but the system goes into oscillations for rather strong perturbations.

In the parameter sense, abrupt excitation of oscillations corresponds to the development of oscillations of a "large" amplitude when a parameter passes through some threshold. Both of these concepts are useful for ecological applications, and it will usually be clear from the context which of them is meant.

Kolmogorov (1936) was the first to point out the possibility of the existence of several nested limit cycles in a predator–prey model. Nevertheless, May (1972, 1973) believed that such a system can only have either a stable equilibrium or a stable limit cycle. Albrecht and his colleagues (Albrecht *et al.*, 1973, 1974) formulated conditions on the types of functions that appear in the predator–prey model in order to ensure a simultaneous existence of stable and unstable cycles, but they did not give any ecological interpretation of those requirements.

In the context of economic models, abruptly excited oscillations were found by Watt (1968) for the system analysed above as well as by Hastings (1981) for a system that hardly admits an ecological interpretation.

The main outcome of the study of system (3.3.4) is that a combination of stabilizing and the destabilizing factors can lead to both gradual and abrupt excitation of oscillations in a predator–prey system.

3.4.3. *Nonlinear Predation at Small Prey Population Density and Predator Saturation (Type III Trophic Function)*

In our analysis of trophic functions for different types of predator–prey interaction, we pointed out that the trophic function for the third type of interaction can be written as

$$b(x) = \frac{bx^2}{1 + A_1 x + A_2 x^2}\,.$$

We make the simplifying assumption that $A_1 = 0$ and consider the system

$$\begin{aligned} \dot{x} &= ax - \frac{bx^2 y}{1 + Ax^2}\,, \\ \dot{y} &= -cy + \frac{dx^2 y}{1 + Ax^2}\,. \end{aligned} \tag{3.4.5}$$

Setting $t = \tau/a$, $x = \sqrt{a/d}u$ and $y = \sqrt{ad/b)}v$, we obtain

$$\begin{aligned} \dot{u} &= u - \frac{u^2 v}{1 + \alpha u^2}\,, \\ \dot{v} &= -\gamma v + \frac{u^2 v}{1 + \alpha u^2}\,, \end{aligned} \tag{3.4.6}$$

where $\gamma = c/a$ and $\alpha = Aa/d$.

The equations for the nullclines of the system have the form

$$\dot{u} = 0, \quad v = \frac{1 + \alpha u^2}{u}\,; \qquad \dot{v} = 0, \quad u = \sqrt{\frac{\gamma}{1 - \alpha\gamma}}$$

For $a\gamma > 1$, there are no nontrivial equilibria in the positive quadrant, so that $\dot{v} < 0$ for all values of variables. In this case, the u-coordinate of the equilibrium, $u_0 = \sqrt{\gamma/(1 - \alpha\gamma)}$, may serve as a parameter, and according to the graphical criterion of stability, the equilibrium is locally stable when $u_0 < 1/\sqrt{\alpha}$ (Fig. 3.4.3*a*). Moreover, constructing the Dulac function (Bautin, Leontovich, 1976) in the form suggested by Hsu (1978) reveals that in this case the system has no limit cycles, that is, the equilibrium is globally stable (see the phase portrait in Fig. 3.3.4). What happens when the parameter u_0 increases through the threshold $1/\sqrt{\alpha}$? Computing the first Lyapunov quantity L_1 does not provide an answer to this question, because in the parameter (u_0, α)-space the quantity L_1 is identically 0 on the Andronov-Hopf curve $u_0 = 1/\sqrt{\alpha}$. (The bifurcation diagram of system (3.4.6) is trivial, consisting of one curve, $u_0 = 1/\sqrt{\alpha}$, and is therefore not shown in a separate figure.) Besides, it has been observed by Khibnik (Bazykin, Khibnik, 1981*b*) that the mentioned Dulac function may be used to prove that system (3.4.6) has no cycles when $u_0 > 1/\sqrt{\alpha}$. As a result, the equilibrium A is globally unstable for $u_0 > 1/\sqrt{\alpha}$, and all trajectories move away toward infinity (as in Fig. 3.3.1). This fact, in turn, implies that the system is conservative on the Andronov-Hopf curve (see Fig. 3.4.3*b*).

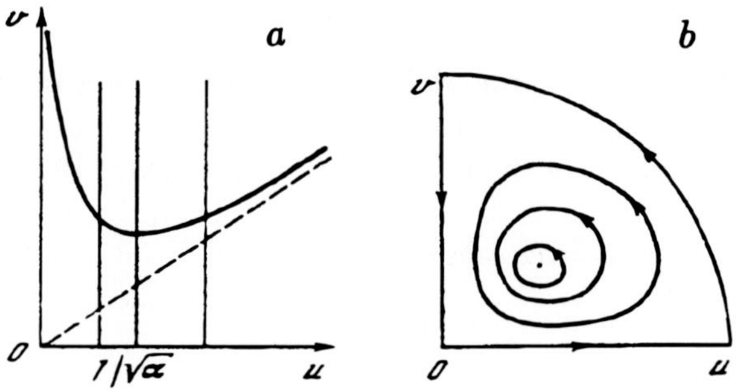

Fig. 3.4.3. Possible relative positions of nullclines (*a*), and phase portrait (*b*) of system (3.4.6) at $u_0 = 1/\sqrt{\alpha}$, when the nullcline $\dot{v} = 0$ passes through the minimum of the nullcline $\dot{u} = 0$.

Thus, we can conclude that, when the parameter u_0 crosses the threshold, the phase portrait changes globally, rather than locally. In physical terms, we can refer to this as an abrupt excitation of oscillations with infinitely large amplitude from a globally stable equilibrium.

We now dwell on the details of these results. We argued above that generically the Andronov-Hopf curve corresponds to a bifurcation of codimension one, and the intersection of this curve in parameter space corresponds to the (dis)appearance of a small limit cycle. In this case, the first Lyapunov quantity L_1 may vanish only at isolated points on the Andronov-Hopf curve. Nevertheless, in a specific parametrization of the model, it may be found that $L_1 \equiv 0$ along the Andronov-Hopf curve. In system (3.4.6) for instance, not only the first Lyapunov quantity, but all Lyapunov qauntities of higher orders, vanish along the entire Andronov-Hopf curve. We say that the bifurcation is of *conditional codimension* one if a single equality constraint on the parameters of a concrete system may actually correspond to a higher degeneracy of the system. In system (3.4.6), the equality $u_0 = 1/\sqrt{\alpha}$ formally corresponds to a bifurcation of infinite codimension.

In fact, we have already met this situation in Sec. 3.3 when we analysed one-parameter modifications of the initial model (3.1.2). Since system (3.1.2) is conservative, every vanishing of the "disturbing" parameter formally corresponds to a bifurcation of infinite codimension. This situation, unlike the one considered in this section, is not of much interest since the corresponding parameters at bifurcation always lie on the boundary of the range of values that make biological sense.

Consider now the interpretation of this result. System (3.4.6) describes the dynamics of predator and prey populations for the third type of functional response of the predator to the prey population density. This type of response simultaneously takes two factors into account. One of them, the quadratic dependence of the predation rate on the density of prey, is a stabilizing factor, whereas the other, predator saturation, is destabilizing. The outcome of the study of Eqs. (3.4.6) is that the dominating role of one of these factors leads to either the global stabilization or the global destabilization of the equilibrium, respectively.

3.4.4. *Predator Competition for Other Resources and Predator Saturation*

The system that simultaneously allows for the stabilizing force of predator competition for resources other than prey and the destabilizing force of predator saturation (type II trophic function) is of the following form (Bazykin, 1974):

$$\dot{x} = ax - bxy/(1 + Ax)\,,$$
$$\dot{y} = -cy + dxy/(1 + Ax) - ey^2\,.$$

Setting $t = \tau/a$, $x = (a/d)u$ and $y(a/b)v$, this system becomes

$$\begin{aligned}\dot{u} &= u - uv/(1 + \alpha u)\,,\\ \dot{v} &= -\gamma v + uv/(1 + \alpha u) - \delta v^2\,,\end{aligned} \tag{3.4.7}$$

where $\gamma = c/a$, $\alpha = Aa/d$ and $\delta = ea/b$.

Following the technique suggested in Sec. 3.4, we first study (3.4.7) at $\gamma = 1$, and then consider the changes of the two-parameter (α, δ)-plane as γ varies. Thus, we consider the system

$$\begin{aligned}\dot{u} &= u - uv/(1 + \alpha u)\,,\\ \dot{v} &= -v + uv/(1 + \alpha u) - \delta v^2\,.\end{aligned} \tag{3.4.8}$$

The equations for the nullclines have the form

$$\dot{u} = 0, \quad v = 1 + \alpha u; \qquad \dot{v} = 0, \quad v = \frac{(1-\alpha)u - 1}{\delta(1 + \alpha u)}\,.$$

If $\alpha > 1$ then $\dot{v} < 0$ over the entire range of variables. The behavior of this system has already been described and interpreted in Sec. 3.3.2 (Eq. (3.3.6)). However, for $\alpha < 1$ there are either no equilibria in the first quadrant or there is a pair A and C of them (Fig. 3.4.4*a*).

The graphical criterion we used so far to determine the local stability of a nontrivial equilibrium cannot be applied to system (3.4.8). Besides, this is the first of the systems we encountered that exhibits more than one nontrivial equilibrium.

Linearizing system (3.4.8) at the equilibria shows that C is a saddle for all values of the parameters and A is either a node or a focus. Depending on the parameters, A may be stable or unstable.

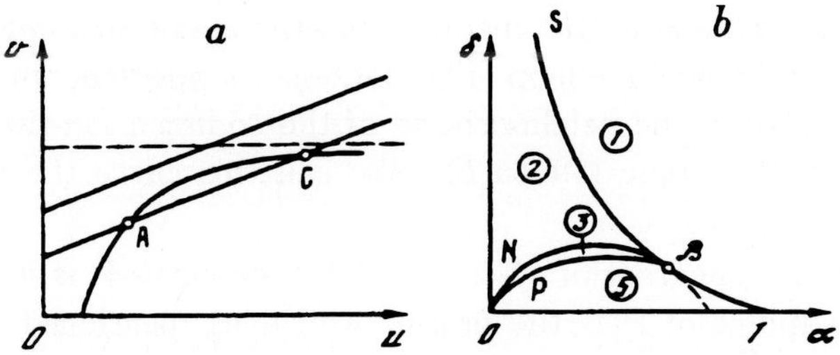

Fig. 3.4.4. Two possible relative positions of nullclines (a), and the bifurcation diagram (b) of system (3.4.8).

The nullclines have a tangency if

$$\delta = (1 - \alpha)^2 / 4\alpha \, . \tag{3.4.9}$$

This equation describes the curve of saddle-node bifurcations dividing the (α, δ)-plane of parameters into regions in which there are either no equilibria or a pair of them (Fig. 3.4.4b) Thus, equation (3.4.9) also provides a condition for the existence of a degenerate equilibrium, namely a saddle-node. We denote the saddle-node curve by S.

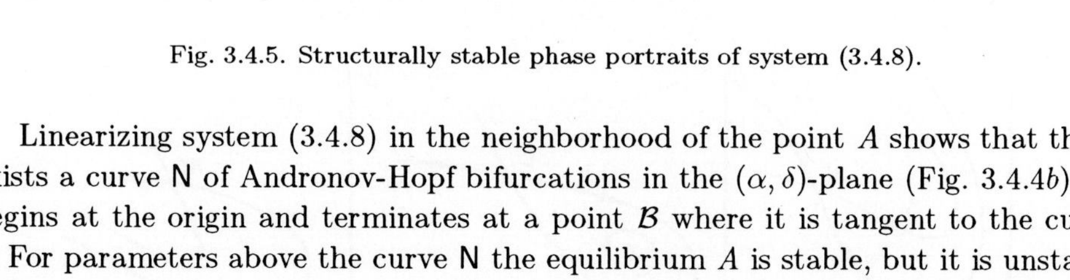

Fig. 3.4.5. Structurally stable phase portraits of system (3.4.8).

Linearizing system (3.4.8) in the neighborhood of the point A shows that there exists a curve N of Andronov-Hopf bifurcations in the (α, δ)-plane (Fig. 3.4.4b). It begins at the origin and terminates at a point $\mathcal{B}$ where it is tangent to the curve S. For parameters above the curve N the equilibrium A is stable, but it is unstable otherwise. The first Lyapunov quantity L_1 is less than zero along the entire curve N. Thus, if the curve N is crossed from above the equilibrium A loses its stability and gives rise to a small stable limit cycle (Fig. 3.4.5).

The point $\mathcal{B}$, at which the curves S and N touch, corresponds to a Bogdanov-Takens bifurcation, that is, to the codimension-two bifurcation where an equilibrium has two zero eigenvalues. This can also be seen as a degenerate saddle-node. The neighborhood in parameter space of that point and the relevant phase portraits were studied in the generic case by Bogdanov (1976*a*, *b*). He proved the existence of exactly one extra bifurcation curve entering $\mathcal{B}$, the contact point of the curves N and S. Along this extra curve the unstable and the stable manifolds of the saddle coincide and form a homoclinic loop (Fig. 3.4.6*b*). Its position, with respect to the Andronov-Hopf curve in the neighborhood of the codimension-two point has also been described by Bogdanov (1976*a*, *b*). We call this curve the homoclinic loop curve an denote it by P.

In the bifurcation diagram of system (3.4.8), the curve P is located below the curve N. One endpoint of P is the origin, which, as mentioned above, formally corresponds to a bifurcation of infinitely large codimension.

These curves in the (α, δ)-plane form the complete bifurcation diagram of the system, since it is not difficult to prove that they have the same relative positions for all values of γ. Note that all four regions have the common boundary point $\mathcal{B}$. The bifurcations that occur in the system may therefore be described by considering a small complete circle around $\mathcal{B}$. We start in region *1* and move counter-clockwise around $\mathcal{B}$ (Fig. 3.4.4*b*).

We have already shown that in region *1*, there is no equilibrium in the first quadrant, and all trajectories go to infinity.

When intersecting the curve S above $\mathcal{B}$ and moving into region *2*, a stable saddle-node AC appears in the phase portrait (Fig. 3.4.6*a*) and splits into a stable node A and a saddle C.

If we keep moving counter-clockwise in the parameter plane, we intersect a curve that is not shown in the bifurcation diagram. When crossing this curve the equilibrium A changes from node to focus, but this is not a topological change of the phase portrait (Fig. 3.4.5, region *2*).

It can be seen that the stable manifold of the saddle C bounds the basin of attraction of the stable equilibrium A, thereby separating two regions of initial conditions. In one region, the to A, and in the other they go to infinity.

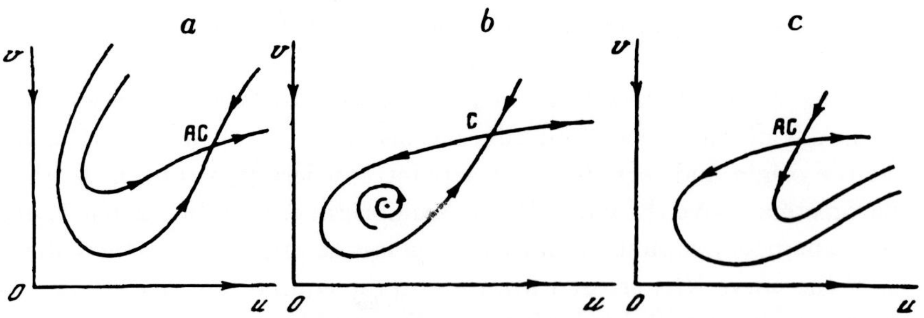

Fig. 3.4.6. Phase portraits of system (3.4.8) for parameters on the bifurcation curves S (above and to the left of point $\mathcal{B}$)(*a*), P (*b*), and S (below and to the right of point $\mathcal{B}$) (*c*).

The next bifurcation occurs when the parameter passes into region *3* by intersecting the Andronov-Hopf curve N. The focus A becomes unstable, and as previously mentioned, a small stable limit cycle appears around it (Fig. 3.4.5, region *3*). In region *3*, the stable manifold of the saddle C bounds the basin of attraction of the stable limit cycle, and, depending on the initial values, the trajectories of the system either accumulate on the cycle or go to infinity.

Moving the parameter further around $\mathcal{B}$, the cycle grows until it finally forms a homoclinic loop with the saddle point C when the parameters hits the curve P (Fig. 3.4.6*b*). After that, the parameters move into region *5*, and the homoclinic loop breaks up (Fig. 3.4.5, region *5*).

The changes of the phase portrait that occur when the curve P is intersected are of fundamental importance. In particular, for parameters from region *3*, there is a large region of initial values in the phase portrait for which the trajectories remain in a finite part of the phase plane for arbitrarily long time while accumulating on the stable limit cycle. However, for parameters in region *5*, all trajectories of the system go off to infinity.

Further events when moving the parameter along in this direction are of little importance, because the rearrangements of the phase portrait do not affect this principle property that all trajectories of the system go to infinity.

This completes the description of the bifurcation diagram and the phase portraits. Although the relative position of the curves N and P is, a priori, unknown, it seems obvious that the homoclinic loop curve exists and comes into the Bogdanov-Takens point $\mathcal{B}$. If we could move the parameters around $\mathcal{B}$ without intersecting the homoclinic loop curve P, it would be impossible to pass from the phase portrait in region *2* to the one in the region *5*.

Ecological interpretation. Depending on the parameters and on the initial condition, the following three regimes are possible within the system:

1. the prey population reproduces without bound, while the predator population density remains asymptotically stable at the level $v = (1/ - \alpha)/\alpha\delta$,
2. the predator and prey populations coexist in a stationary regime (equilibrium A), and
3. the predator and prey populations coexist in a stable oscillatory regime, corresponding to a stable limit cycle.

For parameters from regions *1* and *5* the system's behavior is independent of the initial condition, and the prey population always grows without bound. For parameters from regions *2* and *3*, the phase space of the system splits into two parts: the basin of attraction of a stable equilibrium or stable limit cycle, and the basin of attraction of an infinitely remote point, which corresponds to unbounded reproduction of the prey population.

It is natural to interpret the region of initial values for which the system remains in a finite part of the phase plane for an arbitrarily long time, as the region of normal functioning of the ecosystem, and the boundary of this region in phase space, as "dangerous" for normal ecosystem functioning.

The following example illustrates this interpretation. Assume that u and v are the respective population densities of an insect forest pest and of insectivorous birds that control the pest insects. Then the coexistence of both populations in a stationary or oscillatory state can be regarded as normal functioning of the ecosystem, whereas the unlimited growth of the pest population is considered as a catastrophe. Of course, no pest can reproduce without bound under real biological conditions. However, in the previous model we assume they can since we would otherwise have to take some additional factors into account, which we will not do until later.

Thus, within the framework of the considered model, we can interpret the boundary of the basin of attraction of a stable equilibrium or stable cycle as the phase boundary of the region where the ecosystem functions normally. Therefore, it is interesting to analyse what kind of phase perturbations can drive the system out of this region (Fig. 3.4.7).

To make this clear, we use the example above. Assume that the populations of the insect pest and of the insectivorous birds are in equilibrium. Then, it is intuitively clear that any isolated sharp decrease in the number of birds, from hunting, poisoning, etc., may lead to a catastrophe. It is also evident that an excessive increase in the number of pest animals from immigration or some other phenomenon may also lead to catastrophic consequences.

The model can also predict less obvious phenomena. Suppose for instance that the equilibrium population size of pest insects is unduly large, and we decide to sharply reduce it at a certain moment with massive efforts, such as an application of insecticides. The model can account for the catastrophic consequences of such an action, as depicted in Fig. 3.4.7a by the arrow directed to the left with the dotted trajectory leading out of it. We observe a counter-intuitive effect: the number of pest insects drops sharply, resulting in the death of the birds that feed on them. Then the surviving pest population reproduces and escapes the control by the weakened predator population.

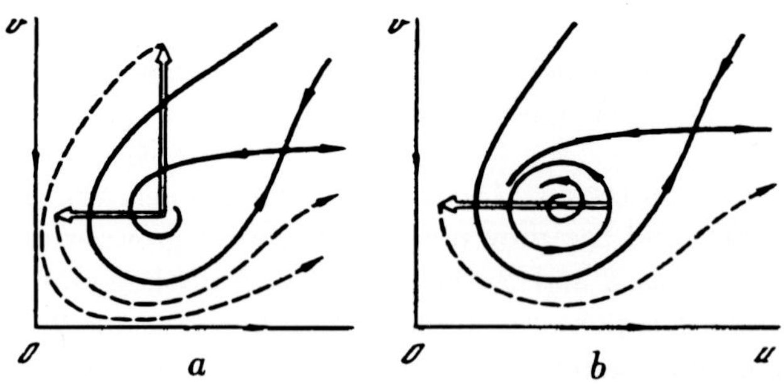

Fig. 3.4.7. When predator and prey exist in stationary (a) and oscillatory (b) states, an abruptly decreased prey population or an increased predator population lead to unlimited growth of the prey population.

Because of its ultimate simplicity, this model cannot claim to describe any real ecological situation. It only demonstrates that counter-intuitive effects are possible in real ecosystems. It is interesting that analogous behavior has been independently described in a model related to the completely different phenomenon of immunity to disease-producing micro-organisms (Molchanov *et al.*, 1971). In conclusion, it should be said that excessive attempts to reduce the pest population may lead to similar catastrophic consequences even if the oscillatory state is regarded as a normal state, as in a regime of permanent outbreaks (Nedorezov,1979; see Fig. 3.4.7*b*).

It is natural to think of the boundary of the basin of attraction of an equilibrium or a stable limit cycle as a phase boundary for the normal functioning of the ecosystem, which is called a *dangerous phase boundary*. Therefore, it also seems natural to single out a region in parameter space where the mentioned regime, or a set of regimes, occurs. The boundary of this parameter region is known as a *dangerous parameter boundary* of the system. Let us consider the bifurcation diagram of system (3.4.8) from this point of view (see Fig. 3.4.4*b*).

The dangerous parameter boundary is a set of bifurcation curves bounding a domain of parameters for which stable equilibria or stable limit cycles exist. For system (3.4.8) this consists of regions *2* and *3*. The dangerous parameter boundary consists of two tangent curves, namely of a part of the saddle-node curve S above the point $\mathcal{B}$ and of the homoclinic loop curve for the saddle C.

From Fig. 3.4.4*b* we can see that for all values of the parameter δ, it is dangerous to either increase or decrease this parameter, although the phenomena that then occur are very different. As δ is increased, the equilibrium remains stable and it approaches the boundary of its basin of attraction until, at the threshold of δ when the curve S is intersected, it reaches the boundary of region *2*. At this point, a saddle-node forms, and the equilibrium finally disappears. By decreasing the parameter δ, the equilibrium first becomes unstable and then gives rise to a small stable limit cycle (gradual excitation of oscillations). This cycle grows and takes on a characteristic shape with a sharp bend near the saddle point. As a consequence, the amplitude increases and the oscillation changes to relaxation type. Finally, the cycle disappears when it becomes a homoclinic loop.

What are the ecological implications of these phenomena? High values of δ mean strong intraspecies competition among predators for resources other than prey. In other words, this means a decrease in the limiting predator population density $v_{\max} = (1 - \alpha)/\alpha\delta$, which is reached when prey is abundant, and it corresponds to a reduction in the capacity of the predator's ecological niche. For large enough values of δ, this capacity becomes so small that even the maximum possible predator population density proves to be insufficient to control the prey. The prey population escapes the predator regardless of its initial density.

On the other hand, with a decrease in δ the limitation of the predator to resources other than prey becomes less pronounced. Here, the destabilizing factor of predator saturation becomes dominant, and the equilibrium becomes locally and then globally unstable as δ decreases.

3.4.5. *Predator Competition for Prey and Predator Saturation*

Considering the destabilizing force of predator saturation and the stabilizing force of competition for prey leads to the following system (Bazykin *et al.*, 1981):

$$\dot{x} = ax - \frac{bxy}{(1+Ax)(1+By)},$$
$$\dot{y} = -cy + \frac{dxy}{(1+Ax)(1+By)}. \qquad (3.4.10)$$

Setting $t = \tau/a$, $x = (a/d)u$ and $y(a/b)v$, this system becomes

$$\dot{u} = u - \frac{uv}{(1+\alpha u)(1+\beta v)},$$
$$\dot{v} = -\gamma v + \frac{uv}{(1+\alpha u)(1+\beta v)}, \qquad (3.4.11)$$

where $\gamma = c/a, \alpha = Aa/d$ and $\beta = aB/b$.

We construct the three-dimensional parameter (α, β, γ)-space of the system in the form of a one-parameter family of two-dimensional (α, β)-planes for different fixed values of the third parameter γ. For $\gamma = 1$, system (3.4.11) is

$$\dot{u} = u - \frac{uv}{(1+\alpha u)(1+\beta v)},$$
$$\dot{v} = -v + \frac{uv}{(1+\alpha u)(1+\beta v)}, \qquad (3.4.12)$$

and the equations for the nullclines are of the form

$$\dot{u} = 0, \qquad u = \frac{1}{\alpha}\frac{(1-\beta)v - 1}{1+\beta v},$$
$$\dot{v} = 0, \qquad v = \frac{1}{\beta}\frac{(1-\alpha)u - 1}{1+\alpha u}. \qquad (3.4.14)$$

Either the nullclines do not intersect or they intersect in two points, corresponding to equilibria A and C (Fig. 3.4.8*a*).

Linearizing equation (3.4.11) at the equilibria shows that, for all values of parameters, C is a saddle and A is either a node or a focus. The equilibria A and C coincide in a saddle-node bifurcation when the the nullclines become tangent.

For system (3.4.12) the equilibrium A is neutral if $\alpha = \beta$. F. S. Berezovskaya and Yu. A. Kuznetsov (Bazykin *et al.*, 1981) showed that system (3.4.12) is Hamiltonian for $\alpha = \beta$, whereas there exist no closed trajectories for $\alpha \neq \beta$. In particular, this implies that the Andronov-Hopf curve N of the equilibrium A coincides with a curve P of homoclinic loops of the saddle C in the parameter (α, β)-plane. Thus, the bifurcation diagram of (3.4.12) is completely described by the two bifurcation curves S and N, where the system undergoes a saddle-node bifurcation and is Hamiltonian, respectively (Fig. 3.4.8*b* and *c*).

The following consideration explains why system (3.4.12) is conservative at $\alpha = \beta$: replacing u by v and vice versa and reversing time gives again the same system.

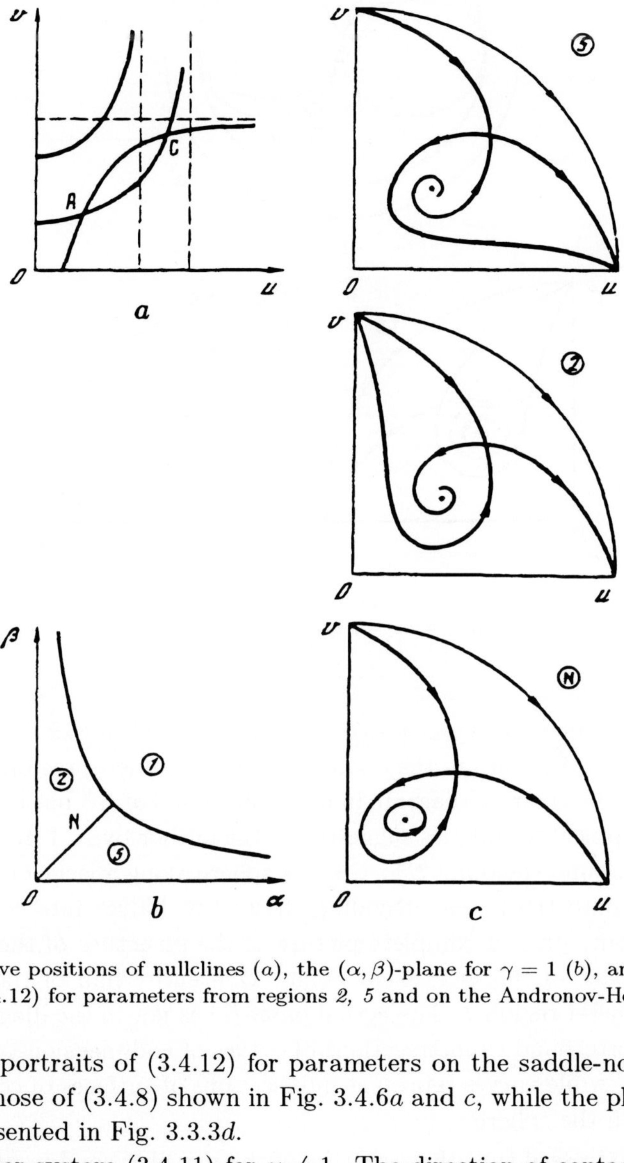

Fig. 3.4.8. Relative positions of nullclines (a), the (α, β)-plane for $\gamma = 1$ (b), and phase portraits (c) of system (3.4.12) for parameters from regions *2*, *5* and on the Andronov-Hopf curve N.

The phase portraits of (3.4.12) for parameters on the saddle-node curve S are analogous to those of (3.4.8) shown in Fig. 3.4.6a and c, while the phase portrait in region *1* is presented in Fig. 3.3.3d.

Now consider system (3.4.11) for $\gamma \neq 1$. The direction of contact of the curves S and N depends on the sign of $\gamma - 1$. Their common point $\mathcal{B}$ in the bifurcation diagram is a Bogdanov-Takens bifurcation point, for which there is an equilibrium with two zero eigenvalues. For $\gamma < 1$ the (α, β)-plane of system (3.4.11) and the (α, δ)-plane of (3.4.8) are qualitatively the same. For $\gamma > 1$ one is the reflection of the other in the diagonal of the first quadrant of the (α, β)-plane (see Fig. 3.4.4b).

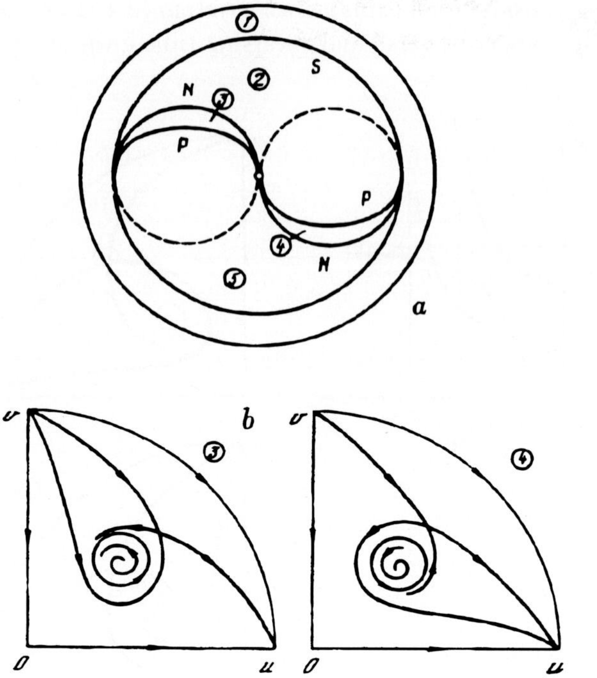

Fig. 3.4.9. The structure of a neighborhood in parameter space of the degenerate point $\mathcal{B}^{\pm}$ (*a*), and phase portraits (*b*) of system (3.4.11) for regions *3* and *4*.

Denote by $\mathcal{B}^{\pm}$ the point $\{\alpha = 1/4, \beta = 1/4$ and $\gamma = 1\}$ in (α, β, γ)-space of maximum degeneracy, which lies in the boundary of all parameter regions of the system (Fig. 3.4.9). The bifurcation diagram in a neighborhood of $\mathcal{B}^{\pm}$ has been constructed by means of graphical techniques suggested to the author by A. I. Khibnik. Imagine a small sphere around the point $\mathcal{B}^{\pm}$. Cut the sphere along a large circle and project the two halves onto the plane. Dividing theses two halves into regions of different phase portraits gives a complete picture of the structure of the neighborhood in parameter space of the mentioned point. One entire half of the sphere corresponds to parameter region *1*. The actual point $\mathcal{B}^{\pm}$ is not in the diagram. Points in that diagram correspond to intersections of curves of codimension-two bifurcations with the sphere, while curves represent intersections of surfaces of codimension-one bifurcations with the sphere.

It should be stressed that the structure of the neighborhood in parameter space of the point $\mathcal{B}^{\pm}$ of system (3.4.11) is not generic, since this neighborhood contains a curve of parameters for which the system is Hamiltonian.

The first Lyapunov quantity L_1 on the Andronov-Hopf curve is negative for $\gamma < 1$, and positive for $\gamma > 1$. In particular, this implies that for $\gamma > 1$, as the Andronov-Hopf curve is crossed from left to right, the equilibrium A becomes unstable due to the disappearance of a small unstable limit cycle (Fig. 3.4.9*b*).

This concludes the construction of the bifurcation diagram and a complete set of phase portraits for the system. Note that system (3.4.11) is the first of the systems studied in this book that depends qualitatively on the parameter γ.

Let us describe the structure of the bifurcation diagram in (α, β, γ)-space in greater detail. It is defined by the relative position of the three surfaces S, N and P of codimension-one bifurcations, corresponding to a saddle-node AC, an Andronov-Hopf bifurcation at A, and a homoclinic loop of the saddle C, respectively. The curve $\mathcal{B}$, which is a curve of contact of the surfaces S, N and P, corresponds to the Bogdanov-Takens bifurcation (of codimension two), where there is an equilibrium with two zero eigenvalues. The curve NP, where the bifurcation surfaces N and P intersect, should correspond, in the generic case, to bifurcations of codimension "one plus one." This means that, for parameters satisfying two equality conditions, two objects in the phase portrait are subject to bifurcations of codimension one: A undergoes an Andronov-Hopf bifurcation and at the saddle C there is a homoclinic loop. Generically, the first Lyapunov quantity may take any value on the curve NP, and vanishes only at certain points. In the concrete case of system (3.4.11), it is identically zero on the whole of NP given by $\gamma = 1$ and $\alpha = \beta$. All Lyapunov quantities of higher orders are also identically zero on this curve. This means that the curve NP, corresponding to a conditional codimension "one plus one", is a curve of conservative systems for (3.4.11), that is, it formally corresponds to bifurcations of infinite codimension.

All phase portraits of system (3.4.11), except those in region *4*, are qualitatively as the corresponding phase portraits of system (3.4.7) described in the previous section. This similarity is quite natural, because both systems describe the action of the same factors, predator saturation and competition among predators, with the only difference that the competition is for different resources. System (3.4.11) describes competition for prey, whereas the competition in system (3.4.7) is for resources other than prey. This difference is manifested in region *4* which is present in the bifurcation diagram of (3.4.11), but not in the one of (3.4.7). The phase portrait in this region is depicted in Fig. 3.4.9*b* and shows that the stable equilibrium is surrounded by an unstable limit cycle that bounds the basin of attraction of this equilibrium. Thus, the boundary of the basin of attraction in region *2* is open (formed by the stable manifold of the saddle), but in region *4* this boundary is closed, so that any sufficiently strong perturbation can drive the system away from the attractor.

Consider the bifurcations of system (3.4.11) for $\gamma > 1$ as the parameters pass from region *2* to region *5* via region *4*. When the homoclinic loop curve P, which separates regions *2* and *4*, is crossed the loop gives rise to a large unstable limit cycle. As the parameters keep changing in the same direction, for instance as α grows, the limit cycle starts to shrink. When the parameters cross the Andronov-Hopf curve N the limit cycle shrinks to a point, the equilibrium loses its stability and all trajectories go off to infinity.

Now, consider the structure of the dangerous parameter boundary of system (3.4.11). For $\gamma < 1$, the dangerous boundary in the (α, β)-plane has the same structure as in the (α, δ)-plane of system (3.4.8), and consists of the analogous pieces of curves.

What is the structure of this dangerous parameter boundary for $\gamma > 1$? It is not difficult to see that regions *2* and *4* correspond to the existence of a stable equilibrium A in the first quadrant of the system, whereas regions *1* and *5* correspond to the absence of such an equilibrium. Hence, the complete dangerous boundary for $\gamma > 1$ is the union of the boundary between regions *1* and *2* and the boundary between regions *4* and *5*. Unlike in the case for $\gamma < 1$, the dangerous parameter boundary is a smooth curve in the (α, β)-plane.

Thus, in (α, β, γ)-space the dangerous parameter boundary of system (3.4.11) consists of pieces of the three surfaces of saddle-node, Andronov-Hopf and homoclinic loop bifurcation. The general structure of the boundary is determined by the structure near the point in parameter space common to all three surfaces, which corresponds to a bifurcation of conditional codimension three for system (3.4.11). A problem with similar properties was investigated by Hainzl (1988).

3.4.6. *Nonlinear Predator Reproduction and Prey Competition*

Combining the destabilizing factor of nonlinear predator reproduction and the stabilizing force of prey competition leads to the system (Bazykin *et. al.*, 1980)

$$\dot{x} = ax\frac{K - x}{K} - bxy \,, \qquad \dot{y} = -cy + dx\frac{y^2}{N + y} \,. \tag{3.4.15}$$

Setting $t = \tau/a$, $x = (a/d)u$ and $y = (a/b)v$ transforms the system to

$$\dot{u} = u - uv - \epsilon u^2 \,, \qquad \dot{v} = -\gamma v + \frac{uv^2}{n + v} \,, \tag{3.4.16}$$

where $\gamma = c/a$, $\epsilon = a/dK$ and $n = bN/a$. There are two possible relative positions of nullclines, depending on the parameters (Fig. 3.4.10*a*).

For all values of the parameters, the origin of the phase space of system (3.4.16) is a saddle point, just as it was for all of the previous systems. There also exists an equilibrium B ($u = 1/\epsilon, v = 0$) on the abscissa, which is a stable node for all parameters.

The condition for a tangency of the nullclines

$$R = (1 - \gamma\epsilon)^2 - 4\gamma\epsilon n = 0 \tag{3.4.17}$$

determines a surface in (γ, ϵ, n)-space. It separates the region of parameters with no nontrivial equilibria phase space from the region in which there are two nontrivial equilibria A and C. For parameters on this surface the equilibria merge, forming a saddle-node AC. This saddle-node surface is denoted by S.

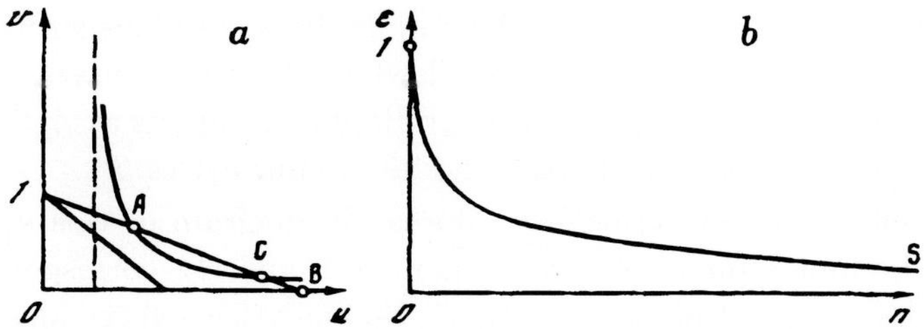

Fig. 3.4.10. Two possible relative positions of nullclines (a), and the curve S of saddle-nodes (b) in the (n, ϵ)-plane for system (3.4.16).

Using the procedure described in the previous sections, we construct the bifurcation diagram in the three-dimensional (γ, ϵ, n)-space of system (3.4.16) in the form of a one-parameter family of bifurcation sets in (ϵ, n)-planes for fixed values of γ. For any γ Eq. (3.4.17) defines a saddle-node bifurcation curve S in the (ϵ, n)-plane, dividing it into two regions (see Fig. 3.4.10b). For parameters above S there are no nontrivial equilibria in the first quadrant of the phase space. For parameters below S there exists a pair of equilibria A and C with coordinates

$$\begin{aligned} u_{\mathbf{A}} &= \frac{1}{2\epsilon}(1+\gamma\epsilon+\sqrt{R}), \quad v_{\mathbf{A}} = \frac{1}{2}(1-\gamma\epsilon-\sqrt{R}), \\ u_{\mathbf{C}} &= \frac{1}{2\epsilon}(1+\gamma\epsilon-\sqrt{R}), \quad v_{\mathbf{C}} = \frac{1}{2}(1-\gamma\epsilon+\sqrt{R}), \end{aligned} \tag{3.4.18}$$

where $R = (1-\gamma\epsilon)^2 - 4\gamma\epsilon n$. The equilibrium C is a saddle for all parameter values, whereas A is either a node or a focus.

Depending on the parameters, A may either be stable or unstable. In the (ϵ, n)-plane (for fixed γ) the condition for the change of stability of A is given by the expression

$$\gamma n/(n+v_A) - \epsilon u_A = 0\,, \tag{3.4.19}$$

which is the equation of the Andronov-Hopf curve N_A for A. As the parameters cross the Andronov-Hopf curve in the direction of increasing n, the equilibrium A loses its stability and the phase portrait changes. Depending on the sign of the first Lyapunov quantity L_1 along N_A, this loss of stability is accompanied by either the birth of a small stable limit cycle T_+, or the disappearance of a small unstable limit cycle T_- shrinking down to A. We shall discuss the sign of L_1 along the Andronov-Hopf curve N_A later.

What other bifurcations of codimension one are possible in this system? There exist two such bifurcations, both of which are nonlocal and occur for parameters that correspond to the appearance of a large limit cycle. We have already come across the bifurcations of this kind. They are the birth of a limit cycle from a homoclinic loop, and the birth of a pair of limit cycles from a saddle-node of limit cycles.

Thus, constructing the bifurcation diagram in the (ϵ, n)-plane for fixed γ requires determining the relative positions of the following bifurcation curves: the curve S of saddle-nodes AC, the curve N_A of Andronov-Hopf bifurcations at A, the homoclinic loop curve P, and the curve Q of saddle-nodes of limit cycles $T_{\pm}$.

The difficulty in constructing the bifurcation diagram of this system is that only two of the four bifurcation curves, namely S and N_A corresponding to local bifurcations, have explicit parametrizations. No analytical expressions are available for the curves P and Q of nonlocal bifurcations, so that one needs to use numerical methods to find their exact positions in the parameter plane with desired accuracy. Within the framework of a qualitative study, however, we are not interested in the exact positions of bifurcation curves. Instead, we are interested in their relative positions, which are locally determined by the structure near points in parameter space where bifurcation curves meet: bifurcation points of codimension two. This local information is sometimes sufficient to completely identify the global structure of the bifurcation diagram. In any case, it makes it possible to drastically reduce the amount of computer calculations needed to find the exact positions of bifurcation curves.

Points of codimension two. The key to constructing the bifurcation diagram of this sytem is to find the bifurcation points of codimension two (or higher) and to study their neighborhoods. This technique proves to be especially effective, because the points corresponding to bifurcations of codimension two are endpoints for some bifurcation curves. If the relative positions of these endpoints are known, we can make conclusions about the relative positions of the bifurcation curves on the whole.

The complete set of curves coming into a point in parameter space of a codimension-two bifurcation, as well as their relative positions and tangencies, are given by the versal unfolding. It describes the neighborhood of a model system of differential equations of the respective singularity (Arnold, 1978). Unfortunately, versal unfoldings are not yet known for all of the codimension-two bifurcations we are interested in.

We now describe all points in the parameter space of system (3.4.16) that correspond to bifurcations of codimension two and are endpoints of bifurcation curves.

1. The bifurcation at the origin O $(\epsilon = 0, n = 0)$ of the parameter plane, corresponding to the initial conservative Volterra system (3.1.2). We have already mentioned that this point is only of conditional codimension two, since it satisfies just two conditions on the system parameters. Formally, this point corresponds to a bifurcation of infinite codimension. This fact is important for us, since in this case, the point O may be the endpoint of any curve of codimension-one bifurcations in system (3.4.16).

2. The point $\mathcal{B}$ of Bogdanov-Takens bifurcations, the intersection point of the curve S of saddle-nodes and the curve N_A of Andronov-Hopf bifurcations at A. We have repeatedly encountered this codimension-two bifurcation earlier, it corresponds to a degenerate saddle-node characterized by two zero eigenvalues. It is important for us that it is the endpoint of two bifurcation curves, N_A and P (Fig. 3.4.11), which are both tangent at the point $\mathcal{B}$ to the curve S. The direction of the curves N_A and P coming into $\mathcal{B}$, and correspondingly, their relative positions in the (ϵ, n)-plane, is determined by the sign of a normal from coefficient, whose vanishing corresponds to an additional degeneracy of the Bogdanov-Takens point $\mathcal{B}$ (Bogdanov, 1976*a*,*b*). If this direction changes as we move along γ then there is a bifurcation of codimension three that will be considered below. We extend the curve N_A beyond the point $\mathcal{B}$ using

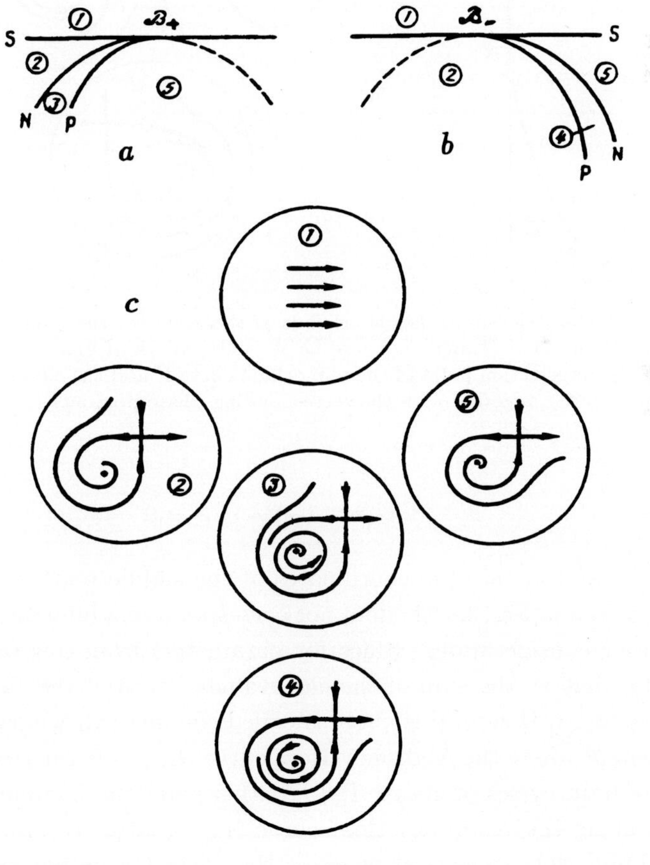

Fig. 3.4.11. Possible structure (*a* and *b*) of a neighborhood of a Bogdanov-Takens point, and the corresponding phase portraits (*c*) after Bogdanov (1976).

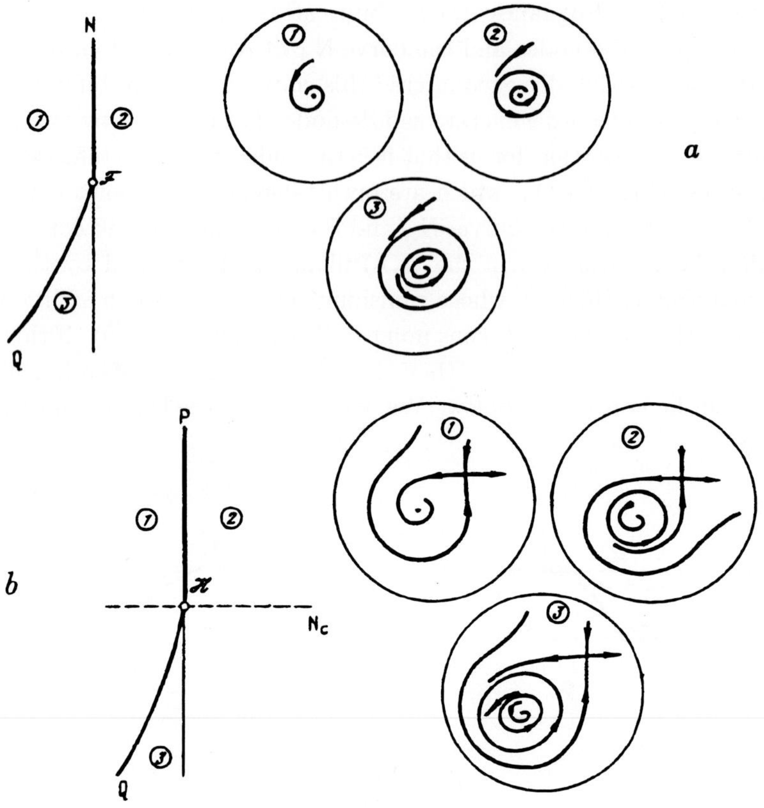

Fig. 3.4.12. The structure of neighborhoods of the point $\mathcal{F}$, the intersection point of the Andronov-Hopf curve N and the curve Q of saddle-nodes of limit cycles (a), and of the point $\mathcal{H}$, the intersection point of the homoclinic curve P and the curve Q of saddle-nodes of limit cycles (b), together with the corresponding phase portraits.

$$\gamma n/(n + v_{\mathbf{C}}) - \epsilon u_{\mathbf{C}} = 0\,,$$

where u_C and v_C are the coordinates of the saddle point C. Although this curve (dotted in Fig. 3.4.11) does not correspond to a bifurcation, it is helpful for further considerations. Since for parameters from this curve the saddle quantity, that is, the sum of the eigenvalues (trace of the Jacobian matrix) of the saddle C is zero, this curve is called the neutrality curve N_C of C.

3. The point $\mathcal{F}$ where the Andronov-Hopf curve N_A meets the curve Q of saddle-nodes of limit cycles (denoted $T_{\pm}$). At this point the Lyapunov quantity L_1 is zero along the Andronov-Hopf curve N_A (Bautin, Leontovich, 1976). A stable limit cycle appears as we cross the curve N_A on one side of the point $\mathcal{F}$, whereas an unstable limit cycle shrinks to a point and disappears on the other side of that point (Fig. 3.4.12a). We have already run into this

codimension-two bifurcation when analysing system (3.4.3). We emphasize that, although there is no parametrization for the curve of saddle-nodes of limit cycles, the location of its end point on the Andronov-Hopf curve N_A can be determined from the condition that the first Lyapunov quantity L_1 be equal to zero on that curve.

4. The point $\mathcal{H}$ where the curve of homoclinic loops meets the curve of saddle-nodes of limit cycles. It corresponds to a neutral homoclinic loop which may generate either stable or unstable limit cycles depending on the direction in which the parameter changes (Fig. 3.4.12*b*). When the parameters cross the homoclinic loop curve a limit cycle appears whose stability is determined by the saddle quantity at the respective saddle. Thus, despite the fact that the neutrality curve N_C of the saddle C is not a bifurcation curve, its point $\mathcal{H}$ of intersection with the homoclinic loop curve P of the saddle C corresponds to a bifurcation of codimension two. This point is the end point for the curve of saddle-nodes of limit cycles (see Fig. 3.4.12*b*). Generically, the structure of the neighborhood in parameter space of a point like $\mathcal{H}$ was studied by Nozdracheva (1982).

Dependence of the (ϵ, n)-plane on γ. The above considerations make it possible to completely determine the qualitative structure of the bifurcation diagram in (γ, ϵ, n)-space of system (3.4.16). We shall describe the structure of this bifurcation diagram in the form of a one-parameter family of two-parameter sections, namely by (ϵ, n)-planes for different fixed values of the parameter γ.

Depending on γ there are three qualitatively different bifurcation sets in the (ϵ, n)-plane. For all values of γ there exist at least three bifurcation curves of codimension-one bifurcations in the (ϵ, n)-plane: the curve S of saddle-nodes AC, the Andronov-Hopf curve N_A, and the homoclinic loop curve P. Furthermore, for all values of γ these curves have the codimension-two Bogdanov-Takens point $\mathcal{B}$ in common.

The structure of the phase portrait for parameters above the curve S in the (ϵ, n)-plane, that is, for parameters from region *1*, does not depend on the value of γ (Figs. 3.4.13 and 3.4.14). In this region, there are no equilibria in the first quadrant, and B is the only attracting equilibrium.

The phase portraits for parameters from regions *2* and *5*, the regions immediately below the curve of saddle-nodes S in the (ϵ, n)-plane, are also independent of the value of γ (Figs. 3.4.13 and 3.4.14).

Thus, the study of the bifurcation diagram in the (ϵ, n)-plane reduces to the construction of the set of possible bifurcations which change the phase portraits in region *2* into those in region *5*.

For $\gamma > 4/3$ the bifurcation diagram in the (ϵ, n)-plane of system (3.4.16) is qualitatively as in the (α, β)-plane of system (3.4.11) for $\gamma > 1$. As shown above, a rearrangement of the bifurcation diagram in the (α, β)-plane of system (3.4.11), takes place as γ passes through the value $\gamma = 1$ at which a curve $\alpha = \beta$ of

Hamiltonian systems appears. This rearrangement is marked by the change in the direction at which the Andronov-Hopf curve N_A and the homoclinic loop curve P comes into the point $\mathcal{B}$.

Let us now examine in detail what happens in the (ϵ, n)-plane of system (3.4.16) when γ passes through the special value $\gamma = 4/3$. As mentioned above, when N_A and P come into the Bogdanov-Takens point $\mathcal{B}$ they are tangent to the saddle-node curve S. When the third parameter γ is changed an additional degeneracy may arise marked by a change in the direction in which the curves N_A and P come into the point $\mathcal{B}$. This is a bifurcation of codimension three, and is precisely what happens in system (3.4.16) at $\gamma = 4/3$.

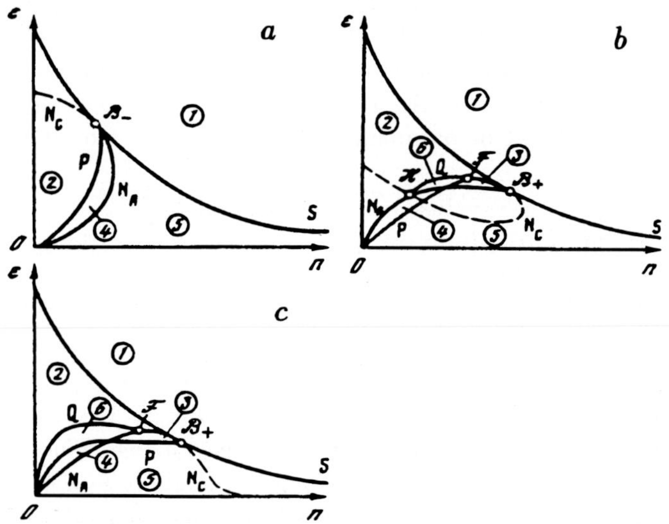

Fig. 3.4.13. The (n, ϵ)-plane as it cuts through the bifurcation diagram in (n, ϵ, γ)-space of system (3.4.16) for $\gamma > 4/3$ (*a*), $4/3 > \gamma > 1$ (*b*), and $\gamma < 1$ (*c*).

Note that changing the direction in which N_A and P come into the point $\mathcal{B}$ also changes their relative positions in a neighborhood of this point. What additional events may this cause?

For $4/3 > \gamma > 1$ the bifurcation diagram of the system has the form shown in Fig. 3.4.13*b*. When a change occurs in the direction of the curves N_A and P coming into $\mathcal{B}$, a point $\mathcal{H}$ appears in the neighborhood of $\mathcal{B}$ at which the curve P of the homoclinic loop intersects the neutrality curve N_C of the saddle C (the latter is represented by a dotted curve in the figure). We know that this point corresponds to a codimension-two bifurcation, namely the existence of a neutral homoclinic loop which, depending on the direction of change of the parameter, may give rise to either a stable or an unstable limit cycle. This bifurcation point is the endpoint of the saddle-node of limit cycle curve Q. The question is where the other end point of the curve Q lies. It is natural to conjecture that the second end point of this curve is the point $\mathcal{F}$ on the Andronov-Hopf curve N_A, where the first Lyapunov quantity L_1 vanishes. A numerical calculation confirms this conjecture and shows that the point in question is located on the Andronov-Hopf curve of the focus A above the point of its intersection with the homoclinic loop curve (Fig. 3.4.13*b*).

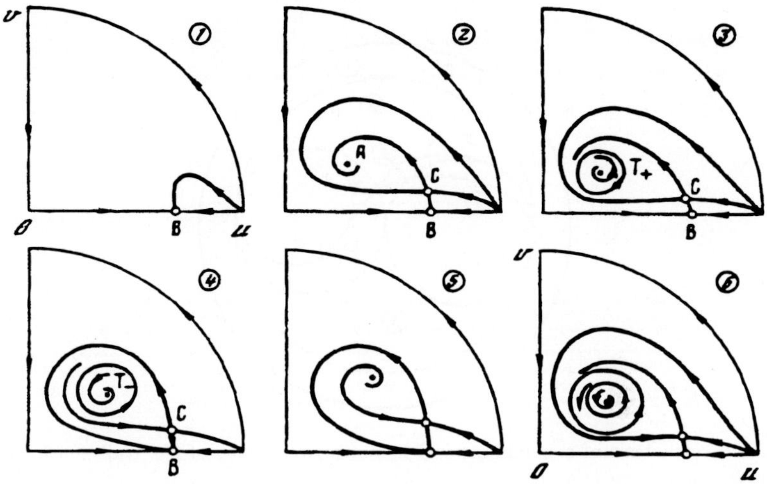

Fig. 3.4.14. Phase portraits of system (3.4.16).

Thus, we conclude that a codimension-three bifurcation that changes the direction of the curves N_A and P (coming into the point $\mathcal{B}$) also leads to the simultaneous appearance in the (ϵ, n)-plane of three points: the two codimension-two bifurcation points $\mathcal{H}$ and $\mathcal{F}$, and the intersection point of N_A and P, corresponding to a bifurcation of codimension "one plus one." These points correspond to the existence of a neutral homoclinic loop, a generalized Andronov-Hopf bifurcation (or doubly degenerate focus), and to a Andronov-Hopf bifurcation simultaneous with a homoclinic loop, respectively. Furthermore, when the parameter γ crosses the special value $\gamma = 4/3$ a new bifurcation curve of codimension one appears in the (ϵ, n)-plane, namely the curve Q of saddle-nodes of limit cycles.

We conjecture that this codimension-three bifurcation, which we just studied in the context of system (3.4.16), represents the generic case. The relative positions of bifurcation surfaces and curves in a neighborhood of this codimension-three point in (γ, ϵ, n)-space is sketched in Fig. 3.4.15.

It is convenient to make use of the graphical technique described earlier, the projection of a sphere around the bifurcation point in parameter space onto a particular plane. The projection of one hemisphere is shown in Fig. 3.4.15. The projection of the other consists entirely of region *1*. The structure shown in Fig. 3.4.15 conjecturally represents the generic case of the codimension-three bifurcation corresponding to a violation of one of the non-degeneracy conditions given by Bogdanov (1976*a*, *b*).

The result of this study serves as an illustration of the ideas of V. I. Arnol'd, who observed that global bifurcations of lower codimensions start playing a role in the analysis of local bifurcations of higher codimension (Arnol'd, 1978).

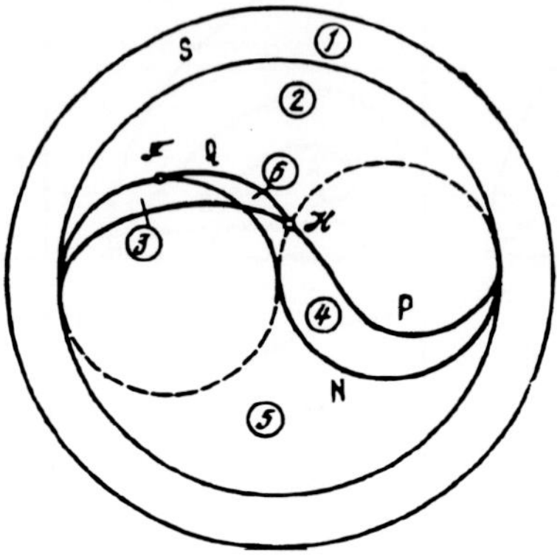

Fig. 3.4.15. The structure of a neighborhood in parameter space of the codimension-three bifurcation point $\mathcal{B}_{\pm}$ in system (3.4.16).

As γ decreases a second rearrangement of the bifurcation diagram in the (ϵ, n)-plane takes place for $\gamma = 1$. The point at which the dashed curve of neutral saddles crosses the ϵ-axis moves toward the origin. For $\gamma = 1$ the curve of neutral saddles undergoes a qualitative rearrangement, and becomes asymptotic to the n-axis for $\gamma < 1$. Correspondingly, as γ decreases from 4/3 to 1, the point $\mathcal{H}$ of neutral homoclinic loops, which is the end point for the curve Q of saddle-nodes of limit cycles, moves along the homoclinic loop curve P from the point $\mathcal{B}$ toward the origin of (ϵ, n)-plane. At $\gamma \leq 1$ the curve of saddle-nodes of limit cycles comes into the origin. Recall that the origin of the (ϵ, n)-plane corresponds to a conservative system and can therefore be the end point of any bifurcation curve.

It should be noted that the rearrangement that takes place in the (ϵ, n)-plane when γ passes through the special value $\gamma = 1$ is not substantial, because it does not lead to either the appearance or the disappearance of regions that correspond to qualitatively new phase portraits. The only difference between the (ϵ, n)-plane for $\gamma \leq 1$ and for $4/3 > \gamma > 1$ is that it is impossible to go directly from region *2* to region *4* by changing γ (see Fig. 3.4.13*b* and *c*).

Now we describe the phase portraits occuring in system (3.4.16) for different values of the parameters. There are only six qualitatively different phase portraits. For parameters from region *1*, the equilibrium B ($u = 1/\epsilon, v = 0$) is the only global attracting equilibrium. Note that this equilibrium exists and is at least locally stable for all other values of the parameters as well.

For parameters from regions *2–6*, the phase portrait of the system contains the two equilibria A and C inside the first quadrant. The equilibrium C is a saddle for all values of parameters, whereas A is either a stable or an unstable node or focus, depending on the parameters. For parameters from region *5*, B is the only globally attracting equilibrium in the phase portrait; the same is true for parameters from region *1* (see Fig. 3.4.14).

For parameters from regions *2–4* and *6*, there exist nontrivial attractors, that is, regimes for which no phase variable of the system vanishes. In fact, the equilibrium A (regions *2*, *4*, and 6) or the stable limit cycle T_+ around A (regions *3* and *6*) are the attractors. The basin of attraction of A is either bounded by the unstable limit cycle T_- (regions *4* and *6*) or by the stable manifold of the saddle C (region *2*). The basin of attraction of the stable limit cycle is always bounded from the outside by the stable manifold of C (regions *3* and *6*), and in addition, it may be bounded on the inside by an unstable limit cycle (region *6*).

The behavior of the system for parameters in region *6* requires more detailed considerations. In this region, for the first time in our study, three attractors exist simultaneously: the nontrivial equilibrium A, the stable limit cycle T_+, and the "semi-trivial" equilibrium B. Of course, these attractors divide the phase space into three basins of attraction. Their boundaries are the stable manifold of the saddle C and the unstable limit cycle T_- surrounding A and lying inside the stable limit cycle T_+.

Ecological interpretation. The fact that the equilibrium B, which corresponds to predator extinction, exists and is locally stable for all parameter values is readily understood. Allowing for competition among prey restricts the prey population's growth even in the absence of the predator. Allowing for the nonlinear reproduction of predators or, in other words, for a delay in the predator population's growth for small predator populations, leads to the existence of a lower critical predator population density. Of course, the initial density of the predator population below which the population is doomed to die out depends on the initial prey density.

For parameters from regions *2*, *4*, and *6* a stable stationary regime of coexistence of the predator a nd prey populations is possible. Coexistence in an oscillatory regime is possible in regions *3* and *6*. For parameters in the narrow region *6*, the predator and prey populations can coexist in either stationary or oscillatory regimes depending on the initial state of the system. We have already run into this phenomenon when studying system (3.4.3). In this case, sufficiently large perturbations of a system which is in the stationary regime always drive it into a regime of stable oscillations. In other words, for parameters from region *6* a regime of abrupt excitation of oscillations in the phase sense may be realised in the system.

To complete the description of the phase portraits, we draw attention to a property that is common to all phase portraits. The basin of attraction of the stable equilibrium may either be bounded by an unstable limit cycle or by the stable manifold of the saddle point. Obviously, in the first case any sufficiently strong perturbation can drive the system beyond the boundary of the basin of attraction of the equilibrium. What about the second case?

It follows from the study of equilibria at infinity that even when the basin of attraction is bounded by the stable manifold of the saddle, almost any sufficiently large perturbation drives the system out of the basin of attraction. In particular, this is a result of the fact that an infinitely far point at the "end" of the abscissa is an unstable node. The only exception is a small class of perturbations for which the predator population density decreases and the prey population simultaneously grows in such a way that the position in phase space remains within a sector formed by the stable manifolds of the saddle point coming from the infinitely far node point (see Fig. 3.4.14, regions *2, 3*, and *6*).

We now explain the fact that almost any sufficiently large perturbation drives the system out of the basin of attraction of the stable state where the predator and prey populations coexist. Contrary to intuition, the model predicts that a sufficiently large increase in predator population density should lead to its eventual extinction. Nonetheless, this result admits the following ecological interpretation: an abrupt increase of the predator population leads to a decrease in the prey population. If the prey population is driven below a level that can sustain the predator, the predator decreases in numbers and may, in turn, drop below a critical level and subsequently die out. Note that, as always in such cases, this is only a verbal description of the model's results and does not serve as an explanation of the phenomenon.

Let us now consider what events occur when the parameters are subject to gradual changes. The most informative (ϵ, n)-plane cutting through the bifurcation diagram is that for $4/3 > \gamma > 1$. This cross section exhibits all parameter regions of the system, making it very convenient to examine what happens as we move along a path in parameter space.

Suppose the parameters are initially from region *2*, and the system itself is in the stable equilibrium A. The first question is what happens if the parameter ϵ is either increased or decreased. It is easy to see that, as long as the parameters remain in region *2*, a gradual increase of ϵ results in a decrease of the equilibrium density of the predator population and in an increase of the equilibrium density of the prey population. This phenomenon can be interpreted as follows: increasing this parameter means decreasing the capacity of the ecological niche of the prey population. In turn, this leads to a decrease in the predator equilibrium density. The equilibrium density of the prey population is governed by both the diminished capacity of its own ecological niche and the pressure of predation. In the present situation, the second of these factors is of greater importance: the decrease of the pressure of predation due to a decrease of the equilibrium density of the predator proves to be more significant than the decrease in the environmental capacity. As a result, the prey population density grows in spite of a decrease in the carrying capacity of its ecological niche.

Regardless of the other system parameters, the system moves to the boundary between parameter regions *2* and *1* with a further increase in ϵ. Crossing this boundary, the stable equilibrium (a node) and the saddle C merge in a saddle-node bifurcation and disappear. Now all trajectories of the system are attracted to the "semi-trivial" equilibrium B, and the nontrivial equilibrium A is abruptly destabilized. To interpret this result, consider what happens if the capacity of the ecological niche of the prey decreases enough so that the equilibrium density of predator drops below a critical density. After that point, the predator population is doomed to die out for any initial densities of the prey and predator populations.

The value of the parameter n is unimportant for the events that occur when ϵ is increased, but it becomes an issue when ϵ is decreased. Consider what happens in this latter case, assuming as before that parameters are from region *2* and the system is in the equilibrium A.

As long as the parameters remain in region *2*, the events also develop independently on ϵ: as the equilibrium density of the predator population grows, the equilibrium size of the prey population drops, although it remains above a threshold $u_{\text{min}} = \gamma$. An explanation of this effect was given above, in the case of increasing ϵ.

The bifurcation that occurs on the boundary of region *2* when ϵ decreases depend on the parameter n (see Fig. 3.4.13*b*). Let us consider these events for a sequence of increasing, but fixed values of n. For $n_{NP} < n_{\mathcal{H}}$, a decrease in ϵ causes the parameters to reach the boundary of region *4*. At that point the system exhibits a homoclinic loop of the saddle C which, with a further decrease in ϵ, generates a "large" unstable limit cycle. An external observer who only sees the equilibrium A does not notice this, because the equilibrium's stability does not change. When the parameter passes through this bifurcation value, only the boundary of the basin of attraction of the equilibrium changes: in region *2* the boundary is the stable separatrix of the saddle C, whereas in region *4* the unstable limit cycle T_- is the boundary.

As ϵ decreases further, the equilibrium A remains stable as long as the parameters stays in region *4*. However, the basin of attraction of the equilibrium, formed by the unstable limit cycle T_-, gradually shrinks. When the parameters are on the boundary between regions *4* and *5*, the unstable limit cycle T_- shrinks to loses its stability resulting in an abrupt change of the dynamics. The "semi-trivial" equilibrium B, corresponding to the extinction of the predator, remains as the only attractor.

From the viewpoint of an external observer watching the equilibrium A, events, when ϵ decreases, develops in the same way for $n_{NP} > n > n_{\mathcal{H}}$. The creation of a pair of limit cycles in a saddle-node of limit cycle bifurcation, when ϵ passes the boundary between regions *2* and *6*, is unnoticable to an external observer, because A remains stable.

As ϵ decreases for $n_{\mathcal{F}} > n > n_{NP}$, events develop in a more complicated fashion. Again, the formation of a pair of limit cycles when the boundatry between regions *2* and *6* is crossed is not noticed by the external observer following the equilibrium A, because A remains stable in region *6*. However, its basin of attraction, bounded by the unstable limit cycle T_-, shrinks. When ϵ passes through the boundary between regions *6* and *3* the unstable limit cycle T_- shrinks to the equilibrium A, which then loses its stability. At that moment, the system exhibits an abrupt excitation of oscillations, because the trajectories reach the stable limit cycle T_+.

By further decreasing ϵ inside region *3* the limit cycle grows. Finally, at the boundary between regions *3* and *5*, the stable limit cycle breaks up as it becomes the homoclinic loop of the saddle C. The coexistence of predator and prey becomes impossible, and the predator dies out.

The events develop in a similar fashion when ϵ is decreased for $n_{\mathcal{B}} > n > n_{\mathcal{F}}$. The difference is that oscillations in the system are gradually excited as ϵ passes the boundary between regions *2* and *3*. A small stable limit cycle is created from the equilibrium A in the phase portrait after that equilibrim loses its stability. With a further decrease in ϵ, the stable limit cycle grows, just as in the preceding case, and finally, breaks up when it becomes a homoclinic loop of the saddle C.

Now we focus on the structure of the dangerous parameter boundary of system (3.4.16). This boundary should be thought of as a surface in (n, ϵ, γ)-space that separates the parameter values for which a nontrivial stable equilibrium or a stable limit cycle exist from parameter values for which the "semi-trivial" equilibrium (corresponding to the predator's extinction) is globally attracting. This definiton of the boundary seems reasonable, because within the framework of this model, the coexisting pair of predator and prey populations represents an ecosystem that is qualitatively different from an isolated prey population. From this point of view, the difference in coexistence between the stationary and the oscillatory states is only of minor importance.

In this formulation, it is reasonable to consider the dangerous parameter boundary of system (3.4.16) as the set of bifurcation surfaces separating regions *2, 3, 4* and *6* from regions *1* and *5* (Fig. 3.4.13). Let us consider the details of the structure of this dangerous parameter boundary in the cross section of the complete bifurcation set given by the (n, ϵ)-plane for $1 < \gamma < 4/3$.

First, note that both increasing and decreasing the parameter ϵ disrupts the coexistence of predator and prey, although the respective mechanisms are qualitatively different. As ϵ grows and crosses the dangerous boundary, the equilibrium state disappears from the phase portrait. With a decrease in ϵ, the coexistence is perturbed in one of two possible ways depending on the other parameters. The equilibrium may lose its stability abruptly with the subsequent extinction of the predator, or it might lose its stability gradually. In the latter case, the stationary regime of coexistence first changes into an oscillatory one. Then these oscillations develop into relaxation-type oscillations until they disappear, leading to the extinction of the predator.

Thus, for a fixed value $1 < \gamma < 4/3$, the dangerous parameter boundary of system (3.4.16) in the (n, ϵ)-plane consists of parts of three different bifurcation curves: the curve S of the saddle-node AC, the Andronov-Hopf curve N_A of the equilibrium A, and the homoclinic loop curve P of the saddle C (see Fig. 3.4.13).

In (n, ϵ, γ)-space the dangerous boundary of this model ecosystem consists of pieces of different bifurcation surfaces of codimension one. These surfaces may intersect pairwise in three ways. First, they may intersect to form "edges", that is, curves corresponding to "one plus one" bifurcations, as in the boundary between regions *3, 4,* and *5.* Second, the surfaces may become tangent to each other and form a cusp, as for example for the boundary between regions *1, 3,* and *5.* Third, two pieces of different surfaces along the dangerous boundary may join each other with a tangency and create a smooth surface, as in the boundary of regions *1, 2* and *4* in Fig. 3.4.13*a*.

The points where all three bifurcation surfaces of the dangerous boundary intersect are codimension-three bifurcation points located at the "corners" of the dangerous boundary. In the case of system (3.4.16) such a point is $\mathcal{B}_{\pm}$, which corresponds to a Bogdanov-Takens point with an additional degeneracy. In the neighborhood of this point the relative positions of the bifurcation surfaces determine the local structure of the dangerous boundary of our model ecosystem. A normal form has not been constructed for this bifurcation, but we conjecture that the relative positions of the bifurcation surfaces that was found for system (3.4.16) and is shown in Fig. 3.4.15 represents the generic case.

Table 2

Stabilizing factors	Destabilizing factors: Saturation of predator	Nonlinear reproduction of prey	Nonlinear reproduction of predator
Competition among prey	(3.4.1)	(3.4.3)	(3.4.6)
Predator competition for prey	(3.4.11)	1	4
Predator competition for other resources	(3.4.7)	2	5
Nonlinearity of predator growth	(3.4.5)	3	6

3.4.7. *Other Two-Factor Modifications*

In the beginning of Sec. 3.4, we set out to study the dynamical features of systems representing two-factor modifications of the Volterra system arising by simultaneously considering one stabilizing and one destabilizing factor. Since we consider four stabilizing and three destabilizing factors, there is a total of twelve systems of this type (see Table 2).

Six of these systems were analysed above in a study that successively presented systems with cenceptually new and more complex features. The combinations of factors numbered 1–6 that appear in the table lead to systems without qualitatively new dynamical features as compared to the systems we did discuss. Therefore, it is not necessary to study them here, because our goal is to reveal common features of systems rather than to analyse concrete ones. Note however, that each of the systems cases 1–6 contains only a single nontrivial equilibrium, and this equilibrium is either stable (locally or globally) or unstable, depending on the specific parameters and the form of the particular system.

3.4.8. *Lower Critical Prey Density*

When we studied the one-factor modifications of the Volterra system (Sec. 3.3.1), we mentioned that allowing for a linear dependence of absolute prey mortality on the population density would not change the form of the initial system (3.1.1). In other words, prey mortality is already taken into account in this system. All of the modifications of the initial system which retain the assumption about the linear reproduction of the prey population should be regarded in a similar way; it may be said that linear prey mortality is automatically taken into account as well. Nevertheless, when the prey population is assumed to reproduce in a nonlinear fashion, allowing for linear prey mortality leads to a qualitatively new effect. In particular, there appears a threshold of the density of the prey population as shown in Section 2.1.

Let us now see what happens to the initial Volterra system (3.1.1) when the two factors of linear prey reproduction at small population densities and linear mortality, are simultaniously taken into account. In other words, we study the dynamics of a predator–prey system in which the prey population satisfy Eq. (2.1.12) in the absence of a predator. Unlike the previous systems of this section that consider one stabilizing and one destabilizing factor, the factor of linear prey mortality cannot be characterized as stabilizing or destabilizing.

We consider the system

$$\begin{aligned} \dot{x} &= ax^2/(N+x) - gx - bxy\,, \\ \dot{y} &= -cy + dxy\,, \end{aligned} \tag{3.4.20}$$

where g is the prey mortality. Setting $t = \tau/a$, $x = Nu$ and $y = (a/b)v$ changes system (3.4.20) into

$$\begin{aligned} \dot{u} &= u^2/(1+u) - \xi u - uv\,, \\ \dot{v} &= -\gamma v + \kappa uv\,, \end{aligned} \tag{3.4.21}$$

where $\xi = g/a$, $\gamma = c/a$ and $\kappa = dN/a$. In this parametrization the value $l = \xi/(1-\xi)$ is the lower threshold density of the prey population.

Depending on the parameters, the relative position of nullclines may be as shown in Fig. 3.4.16*a*, and the phase portraits are as shown in Figs. 3.4.16*b* and *c*.

In contrast to all other systems studied in this section, the equilibrium at the origin is stable for all parameter values. The point B $(u = l, v = 0)$ represents an unstable node if $\gamma/\kappa < l$ and a saddle with a stable manifold in the first quadrant if $\gamma/\kappa > 1$. The nontrivial equilibrium in the first quadrant only exists for $\gamma/\kappa > l$ and it is always unstable.

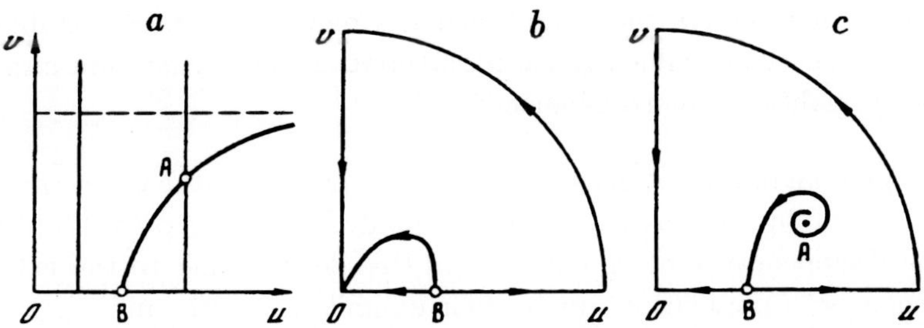

Fig. 3.4.16. Two possible relative positions of nullclines (a), and the corresponding phase portraits (b and c) of system (3.4.21).

The phase portraits shown in Figs. 3.4.16b and c admit a natural interpretation. If γ/κ, the prey population density necessary for the predator population to exist is smaller than the threshold l then there is no equilibrium in the first quadrant, and predator and prey cannot coexist. If the equilibrium density of prey that can sustain the predator population is more than the lower critical density l, then a nontrivial equilibrium, corresponding to the coexistence of predator and prey, exists, although it is unstable. As before, the origin remains the only global attractor. Hence, incorporating the effect of lower critical size of the prey population in the predator–prey system dooms both populations to extinction for all parameter values.

Recall that allowing for only the single factor of nonlinear reproduction of prey (system (3.3.2)) results in the existence of a nontrivial equilibrium for all values of parameters, but it appears to be globally unstable. While spiraling away from this equilibrium, the trajectories of the system wind around an infinitely far limit cycle, that is, the predator and prey population sizes oscillate with an amplitude that grows without bound. Therefore, the result concerning the inevitable extinction of both populations can be interpreted as follows: nonlinear reproduction of the prey population destabilizes the sizes of both populations and makes them oscillate with an ever-increasing amplitude. When the amplitude of oscillations becomes so large that the prey population size drops below threshold then the prey population and, consequently, the predator population dies out.

We have now considered the complete set of two-factor modifications of the Volterra system that take one stabilizing and one destabilizing factor into account. In all cases a complete qualitative study of the corresponding systems of differential equations has been carried out. The models have demonstrated three nontrivial regimes, that is, regimes that do not involve the extinction of either population:

(a) stationary coexistence of both populations,
(b) coexistence of both populations in an oscillatory fashion,

(c) unlimited growth of the prey population or of both populations.

Depending on the species form of the model, each of these regimes may either be globally stable or have a bounded basin of attraction. In the latter case, for a fixed values of parameters the system exhibits one of the three possible regimes depending on the initial value. When parameters change the system may undergo the following qualitative rearrangements:

(a) the gradual or abrupt excitation of oscillations, with the latter characterized by an amplitude that may either be limited or may grow without bound,
(b) the disappearance of an equilibrium that leads either to the extinction of one or both populations, or to their unlimited growth, and
(c) the disappearance of oscillations in a homoclinic loop.

3.5. Three-Factor Modifications of the Volterra Model

Now we introduce the factor of prey competition to the biologically ill-posed models (3.4.5), (3.4.7), (3.4.11) and (3.4.20), thereby making it impossible, in the new models, that the prey population grows without bound, even in the absence of a predator. To achieve this we need to simultaneously allow for three factors affecting the dynamics of predator-prey systems, on top of what is considered in the classical Volterra model. Thus, in this section we shall study a three-parameter perturbation of system (3.1.1), that is, systems of differential equations depending on four parameters.

3.5.1. *Predator Saturation, Nonlinear Predation (Type III Trophic Function) and Competition among Prey*

Taking into account these three factors in the initial model (3.1.1) or, in other words, allowing for prey competition in model (3.4.5), yields

$$\begin{aligned} \dot{x} &= ax - \frac{bx^2y}{1+Ax^2} - ex^2\,, \\ \dot{y} &= -cy + \frac{dx^2y}{1+Ax^2}\,. \end{aligned} \qquad (3.5.1)$$

System (3.5.1) was studied by Kasarinoff and Deiesch (1978), but they did not construct the bifurcaton diagram. Let us study this system completely.

Setting $t = \tau/a$, $x = \sqrt{a/d}u$ and $y = (\sqrt{ad}/b)v$ changes system (3.5.1) into

$$\begin{aligned} \dot{u} &= u - u^2v/(1+\alpha u^2) - \epsilon u^2\,, \\ \dot{v} &= -\gamma v + u^2v/(1+\alpha u^2)\,, \end{aligned} \qquad (3.5.2)$$

where $\gamma = c/a$, $\alpha = Aa\sqrt{a/d}/b$ and $\epsilon = e/\sqrt{ad}$.

It is convenient to present the bifurcation diagram by choosing the parameter α as one axis and $u_0 = \sqrt{\gamma/(1-\alpha\gamma)}$ as the other axis (Fig. 3.5.1).

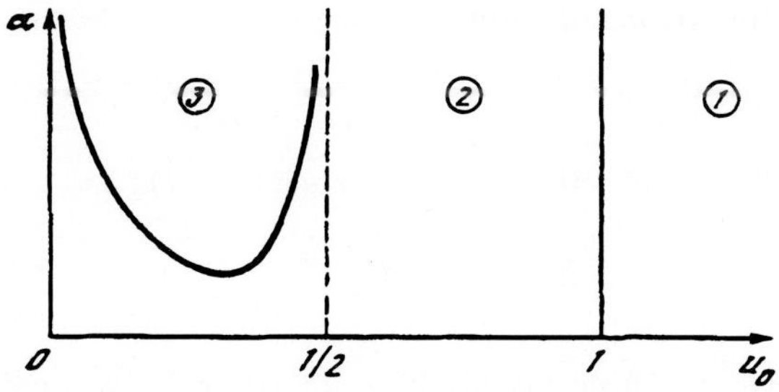

Fig. 3.5.1. The bifurcation diagram of system (3.5.2).

The phase portraits of systems (3.5.2) and (3.4.1) are identical (see Fig. 3.4.1*c*), and their interpretations are similar. There are no qualitatively new effects in system (3.5.2).

3.5.2. *Predator Saturation, Predator Competition for Resources Other than Prey, and Competition among Prey*

Allowing for these three factors, or in other words, including the additional factor of prey competition in model (3.4.7), leads to the system

$$\begin{aligned} \dot{x} &= ax - bxy/(1+Ax) - ex^2\,, \\ \dot{y} &= -cy + dxy/(1+Ax) - hy^2\,. \end{aligned} \tag{3.5.3}$$

This system was proposed by Alexeev (1973) and Bazykin (1974). It was first studied partially, and later completely, by Bazykin and his colleagues (Bazykin, 1974, 1976; Bazykin *et al.*, 1980). Setting $t = \tau/a$, $x = (a/d)u$ and $y(a/b)v$ changes system (3.5.3) into

$$\begin{aligned} \dot{u} &= u - uv/(1+\alpha u) - \epsilon u^2\,, \\ \dot{v} &= -\gamma v + uv/(1+\alpha u) - \delta v^2\,, \end{aligned} \tag{3.5.4}$$

where $\gamma = c/a$, $\alpha = Aa/d$, $\epsilon = e/d$ and $\delta = h/b$.

Thus, the system we are to study depends on four parameters. The task is not too difficult because the behavior of the system for $\delta = 0$ and for $\epsilon = 0$ has already been studied above ((3.4.2) and (3.4.8), respectively).

Let us consider this system. As before, we treat γ as a separate parameter that comes from the unperturbed Volterra system, and we first study (3.5.4) for $\gamma = 1$. Then, after the bifurcation diagram in $(\alpha, \delta, \epsilon)$-space for $\gamma = 1$ is constructed, we intend to follow its changes as γ is varied.

Again, it is reasonable to construct the three-dimensional $(\alpha, \delta, \epsilon)$-space as a one-parameter family of two-dimensional (α, δ)-planes for various fixed values of ϵ.

For $\epsilon = 0$ the bifurcation diagram of system (3.5.4) coincides with that of system (3.4.8). Consider it now for $0 < \epsilon \ll 1$. First we locate the curves corresponding to bifurcations of equilibria in the (α, δ)-plane, and then we do the same for the curves corresponding to bifurcations of limit cycles.

Bifurcations of equilibria. The equations of the nullclines of the system are

$$\begin{aligned} \dot{u} = 0, \quad & v = (1 + \alpha u)(1 - \epsilon u)\,; \\ \dot{v} = 0, \quad & v = \frac{1}{\delta}\left(\frac{u}{1 + \alpha u} - 1\right). \end{aligned} \tag{3.5.5}$$

Generically, there are three possibilities for the relative positions of nullclines: no intersection, one point of intersection (A) in the first quadrant, and three points of intersection (A_1, C and A_2) in the first quadrant, respectively (Fig. 3.5.2).

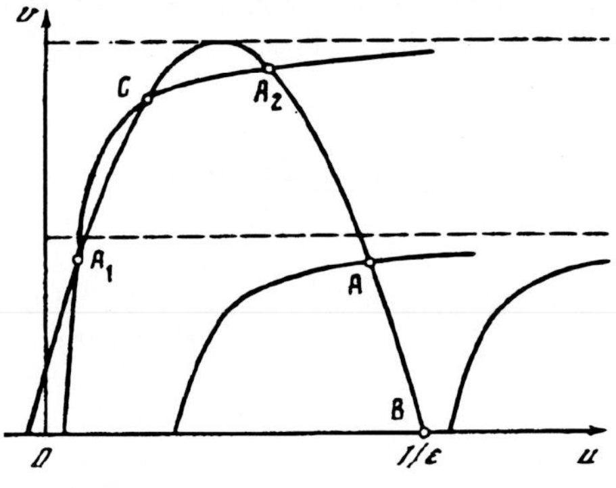

Fig. 3.5.2. Three possible relative positions of nullclines of system (3.5.4) in the generic case.

For all values of the parameters, the origin of the phase plane is a saddle equilibrium, and the point B ($u = 1/\epsilon, v = 0$) is another equilibrium of the system.

The point B is a globally stable node for parameters such that the nullclines do not intersect in the first quadrant. For all other values B is a saddle. The "intermediate" equilibrium C is also a saddle when there are three equilibria in the phase portrait. The equilibria A, A_1 and A_2 are nonsaddles. The type of their stability will be considered below.

The points A_1 and C, as well as A_2 and C can merge. This defines two curves S_1 and S_2 of saddle-node bifurcations in the (α, β)-plane corresponding to the saddle-nodes A_1C and A_2C, respectively.

Analytically, these conditions are given by roots of multiplicity two of the cubic equation

$$\delta(1 + \alpha u)^2(1 - \epsilon u) = (1 - \alpha)u - 1\,. \tag{3.5.6}$$

The saddle-node bifurcation curves in the (α, β)-plane have two common points $\mathcal{A}_1$ and $\mathcal{A}_2$ (Fig. 3.5.3), and they bound a closed crescent-shaped region or *biangle*.

The end points are cusp bifurcation points where two nodes of identical stability and a saddle merge. At the points $\mathcal{A}_1$ and $\mathcal{A}_2$ the curves S_1 and S_2 are tangent, asymptotically they are forming a semi-cubic parabola.

Consequently, the curves S_1 and S_2 of saddle-nodes A_1C and A_2C divide the (α, β)-plane into two regions: parameters inside the biangle conrrespond to the existence of three nontrivial equilibria in the phase portrait, while parameters outside the biangle correspond to the existence of a single nontrivial equilibrium.

As $\epsilon \to 0$, the lower arc of the biangle, corresponding to the birth of the saddle-node A_2C, "gets glued" to the coordinate axes of the (α, δ)-plane. Here, the point $\mathcal{A}_1$ tends to $(\alpha = 0, \delta = \infty)$, and the point $\mathcal{A}_2$ tends to $(\alpha = 1, \delta = 0)$, so that the bifurcation diagram of Fig. 3.5.3 takes the form shown in Fig. 3.4.4*b*.

Let us analyse the stability of the points A, A_1 and A_2. Qualitatively, this can be done without any mathematical computations, by starting with the bifurcation sets of system (3.5.4) for $\epsilon = 0$ and $\delta = 0$ (which are identical to those of systems (3.4.8) and (3.4.2) in Figs. 3.4.4*b* and 3.4.1*b*, respectively) and using continuity.

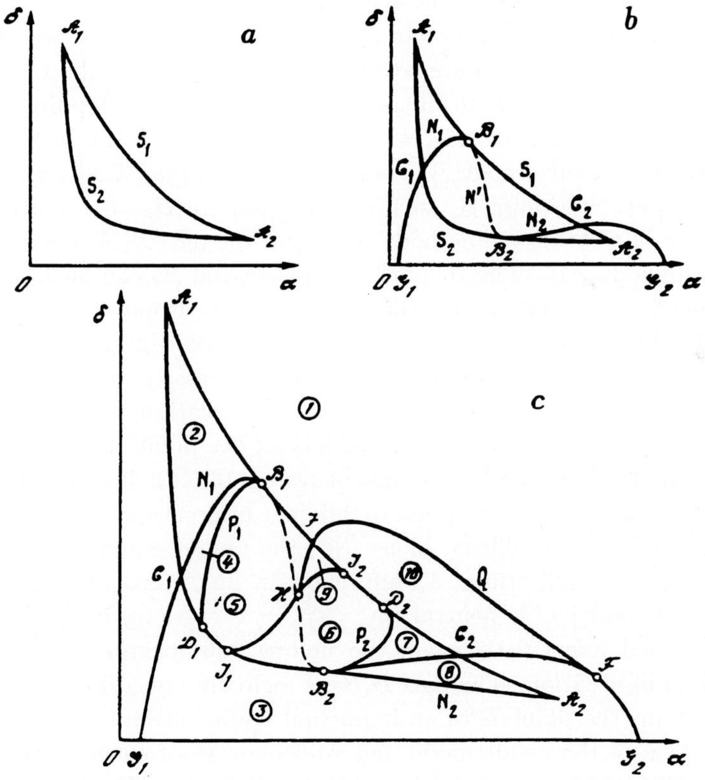

Fig. 3.5.3. Successive stages of constructing the bifurcation diagram of system (3.5.4): the curves of saddle-nodes (*a*), the Andronov-Hopf curve (*b*), the homoclinic curve and the curve of saddle-nodes of limit cycles (*c*) are added.

Consider the location of the Andronov-Hopf curve in the (α, δ)-plane. Comparing it with the bifurcation set (Fig. 3.4.1*b*) of system (3.4.2) shows that the abscissa of the (α, δ)-plane of system (3.5.4) has two common points $\mathcal{G}_1(\alpha = \alpha_1, \delta = 0)$ and $\mathcal{G}_2(\alpha = \alpha_2, \delta = 0)$ with the Andronov-Hopf curve, see Fig. 3.5.3*b*.

Comparing this with the bifurcation diagram of system (3.4.8) in Fig. 3.4.4 shows that the Andronov-Hopf curve, if continued, should touch the upper arc of the biangle at a point $\mathcal{B}_1$ (Fig. 3.5.3*b*). Hence, the Andronov-Hopf curve intersects the lower arc of the biangle at a point $\mathcal{C}_1$ between the points $\mathcal{G}_1$ and $\mathcal{B}_1$. For parameters on the segment $\mathcal{G}_1\mathcal{C}_1$ the unique equilibrium A in the phase portrait loses stability, and for parameters on the segment $\mathcal{C}_1\mathcal{B}_1$ A_1 loses stability. The bifurcation point $\mathcal{C}_1$ corresponds to a codimension "one plus one" bifurcation: the saddle-node bifurcation of C with A_2, simultaneously with the Andronov-Hopf bifurcation of the equilibrium A_1. The Bogdanov-Takens point $\mathcal{B}_1$ corresponds to an equilibrium with two zero eigenvalues, as was discussed earlier (in particular, when studying system (3.4.8)). Here, the part $\mathcal{A}_1\mathcal{B}_1$ of the upper arc of the biangle corresponds to the the existence of a *stable* saddle–node A_1C in the phase portrait, whereas the part $\mathcal{B}_1\mathcal{A}_2$ corresponds to the existence of an *unstable* saddle–node A_1C.

Consider now the position of the Andronov-Hopf curve between the points $\mathcal{B}_1$ and $\mathcal{G}_2$ (Fig. 3.5.3*b*). Since the points $\mathcal{A}_1$ and $\mathcal{A}_2$ correspond to codimension-two bifurcations, they should be common to the curves of saddle-nodes of *identical stability*: the bifurcation point $\mathcal{A}_1$ in the phase portrait is the birth point of the saddle C and two *stable* nodes A_1 and A_2, whereas the point $\mathcal{A}_2$ is the birth point of the saddle C and two *unstable* nodes A_1 and A_2. This implies that on the lower arc of the biangle there is a point $\mathcal{B}_2$ such that the part $\mathcal{A}_1\mathcal{B}_2$ of the arc corresponds to the existence of a stable saddle-node A_2C, and the part $\mathcal{B}_2\mathcal{A}_2$ corresponds to an unstable saddle-node A_2C. The Bogdanov-Takens point $\mathcal{B}_2$ can be seen as a degenerate saddle-node A_2C and is the point where the Andronov-Hopf curve and the lower arc of the biangle touch. How this curve continues is obvious: it intersects the upper arc of the biangle at a point $\mathcal{C}_2$ and hits the abscissa at a point $\mathcal{G}_2$. The point $\mathcal{C}_2$ corresponds to a codimension "one plus one" bifurcation, namely a saddle-node AC simultaneously with the change of stability of the point A_2. Now this curve is constructed completely in the (α, δ)-plane of system (3.5.4). It should be noted that different parts of this curve correspond to different bifurcations. the Andronov-Hopf bifurcations of different equilibria: along $\mathcal{G}_1\mathcal{C}_1$ the unique equilibrium A bifurcates, and along $\mathcal{C}_1\mathcal{B}_1$ the equilibrium A_1 bifurcates in an Andronov-Hopf bifurcation. Along $\mathcal{B}_1\mathcal{B}_2$ the saddle C is neutral ($|\lambda_1^C/\lambda_2^C| = 1$). Along $\mathcal{B}_2\mathcal{C}_2$ the equilibrum A_2 bifurcates, and along $\mathcal{C}_2\mathcal{G}_2$ again the unique equilibrium A bifurcates in an Andronov-Hopf bifurcation. The part $\mathcal{B}_1\mathcal{B}_2$ is formally not a bifurcation curve, although it contains the point $\mathcal{H}$ of an important codimension-two bifurcation. The relative positions of the saddle-node and Andronov-Hopf curves, discussed above by general considerations, were confirmed by computations (Buriev, 1983).

Bifurcations of limit cycles. Apart from the bifurcation curves of equilibria (Fig. 3.5.3*b*), the bifurcation diagram of system (3.5.4) contains curves of bifurcations of limit cycles. We now study the actual bifurcations and the relative positions of the corresponding curves. A curve P_1 emerges from the point $\mathcal{B}_1$ into the interior of the biangle; it corresponds to the existence of a homoclinic loop of the saddle C around the point A_1. This curve has a common point $\mathcal{D}_1$ with the lower arc of the biangle, and, naturally, cannot continue beyond the biangle (since this marks the disappearance of the saddle).

The structure of a neighborhood of a point common to a curve of saddle-nodes and a curve of homoclinic loops was studied by Lukyanov (1982). Although the asymptotics of those curves have not been studied in the neighborhood of the point $\mathcal{D}$, there are reasons to believe that those curves are tangent in the generic case.

In complete analogy, a curve P_2 emerges from point $\mathcal{B}_2$ into the interior of the biangle. It corresponds to the existence of a homoclinic loop of the saddle C around the point A_2. This curve has a common point $\mathcal{D}_2$ with the upper arc of the biangle.

We now construct phase portraits for parameters from different regions in order to see which other curves must be present in the bifurcation diagram.

Figure 3.5.4 shows that the phase portraits for regions *5* and *6* only differ by a large limit cycle surrounding all three equilibria. It is clear that it is the bifurcation af a large homoclinic loop of the saddle C around the equilibria A_1 and A_2 that transforms phase portrait *5* into phase portrait *6*. In other words, the bifurcation diagram of the system should contain a curve corresponding to the existence of this large homoclinic loop. Naturally, this bifurcation curve should lie completely inside the biangle since outside the saddle C is absent. The mentioned curve has two points $\mathcal{I}_1$ and $\mathcal{I}_2$ in common with the arcs of the biangle. As shown in Fig. 3.5.3*c*, the point $\mathcal{I}_1$ is located between the points $\mathcal{B}_2$ and $\mathcal{D}_1$, and the point $\mathcal{I}_2$ between the points $\mathcal{B}_1$ and $\mathcal{D}_2$.

The bifurcation diagram of the system is not yet complete. As is clear from from Fig. 3.5.3*c*, the curve $\mathcal{I}_1\mathcal{I}_2$ of the large homoclinic loop intersects the neutrality curve $\mathcal{B}_1\mathcal{B}_2$ of the saddle C at a point $\mathcal{H}$. This point corresponds to the codimension-two bifurcation of a neutral homoclinic loop, the bifurcation studied by Nozdracheva (1982). This implies the following: if the parameters intersect the curve $\mathcal{I}_1\mathcal{I}_2$ of the large homoclinic loop from above through $\mathcal{I}_1\mathcal{H}$ then the homoclinic loop gives birth to a stable limit cycle. If the parameters intersect the curve $\mathcal{I}_1\mathcal{I}_2$ between $\mathcal{H}$ and $\mathcal{I}_2$ then an unstable limit cycle breaks up in the homoclinic loop. In particular, this indicates that $\mathcal{H}$ is the end point for the curve Q of saddle-nodes of limit cycles.

Consequently, if the parameters pass from region *6* to region *5*, two changes in the phase portrait may occur: either the large stable limit cycle breaks up as it becomes the homoclinic loop, or the homoclinic loop gives birth to an unstable limit cycle inside a stable one (see parameter region and phase portrait *9*). In the latter case both cycles move toward each other as the parameters change further and disappear when the parameters cross the saddle-node of limit cycle curve Q.

Now it remains to find the second end point of the curve Q of saddle-node of limit cycles. It is natural to assume that there is a point $\mathcal{F}$ between the points $\mathcal{C}_2$and $\mathcal{G}_2$ on the Andronov-Hopf curve where the first Lyapunov quantity is equal to zero. This codimension-two bifurcation point is the end point for the curve Q of saddle-nodes of limit cycles. Computations have confirmed this conjecture (Buriev, 1983). Now the bifurcation diagram in the (α, δ)-plane of system (3.5.4) for $\gamma = 1$ is complete (Fig. 3.5.3*c*). The phase portraits for regions *1–10* are sketched in Fig. 3.5.4.

Now we see how the bifurcation diagram in the (α, δ)-plane varies as ϵ is increased. Note that for $\epsilon \ll 1$ the structure of this bifurcation diagram is independent of the value of γ. The (α, δ)-plane for $\epsilon > 1/4$ does not depend on γ either: the whole first quadrant of the (α, δ)-plane is region *1*, because, as follows from the corresponding algebraic equations, the pairs of points $\mathcal{A}_1$, $\mathcal{A}_2$ and $\mathcal{G}_1$, $\mathcal{G}_2$ come closer to each other, merge, and finally disappear as ϵ grows. This happens for any value of the parameter γ.

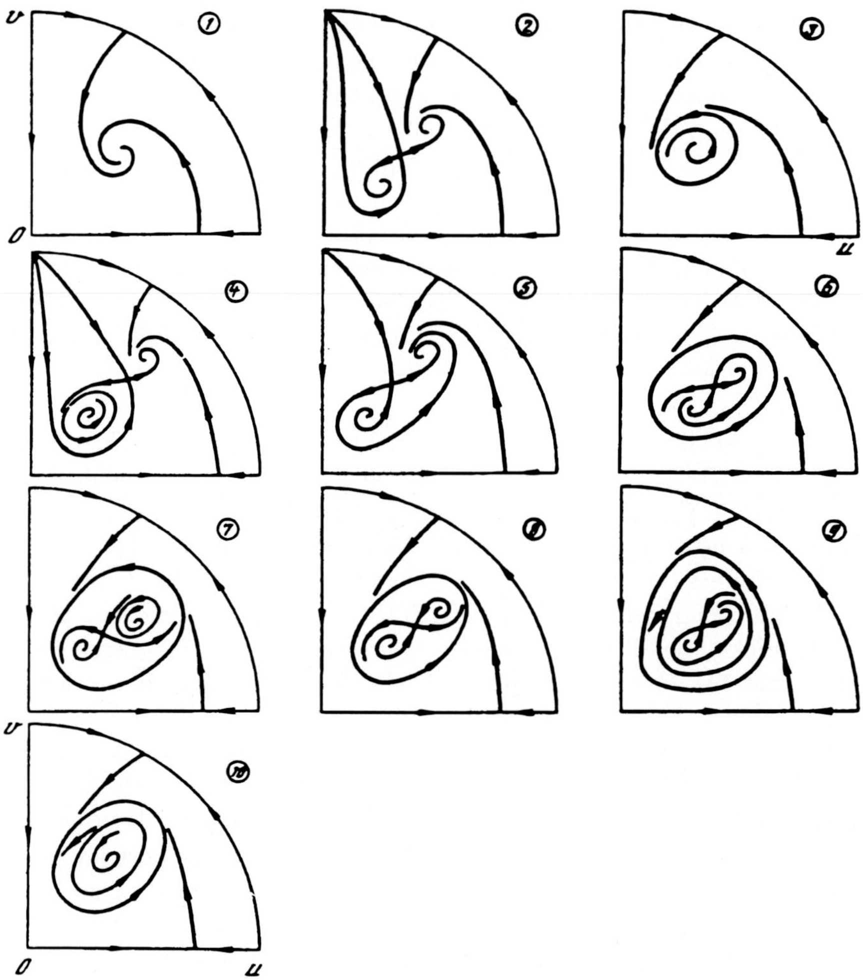

Fig. 3.5.4. The complete set of structurally stable phase portraits of system (3.5.4).

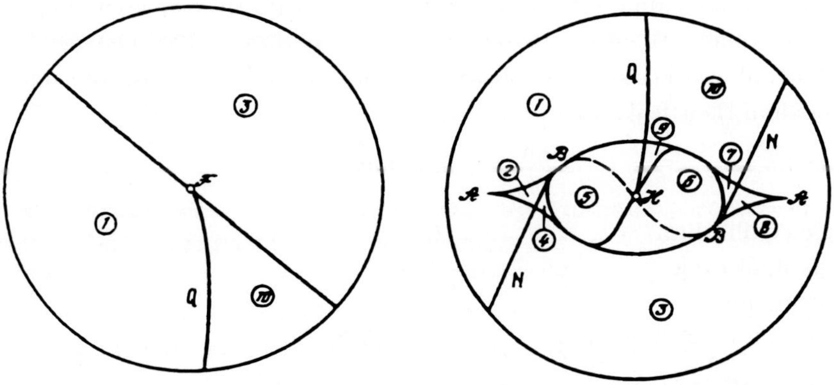

Fig. 3.5.5. The structure of a neighborhood of the codimension-three bifurcation point in $(\alpha, \delta, \epsilon)$-space at an intersection of the cusp bifurcation curve A and the surface of Andronov-Hopf bifurcations N.

What happens to the bifurcation diagram in the (α, δ)-plane before the points $\mathcal{A}_1$ and $\mathcal{A}_2$ merge? It is obvious that generically both points, when ϵ changes, should be either above or below the Andronov-Hopf curve N just before they disappear from the bifurcation diagram. In other words, before its disappearing from the bifurcation diagram the biangle should either completely sink below the curve N or rise above it. The corresponding bifurcations that occur in the (α, δ)-plane were numerically studied by Buriev (1983). The main event is the passing of the Andronov-Hopf curve N through one of the corners of the biangle (at the point $\mathcal{A}_1$ or $\mathcal{A}_2$) for a certain value of ϵ. This is a codimension-three bifurcation that has not been studied in the generic case. The structure of parameter space in a neighborhood of this point for system (3.5.4) has been obtained by means of projecting a sphere around the point onto the plane (Fig. 3.5.5). We conjecture that this represents a generic case of the intersection of a cusp bifurcation curve and a surface of Andronov-Hopf bifurcations in a three-parameter space. The intersection point corresponds to the violation of one of the nondegeneracy conditions for the Bogdanov-Takens bifurcation (Bogdanov 1976*a*,*b*).

In the (α, δ)-plane, increasing ϵ is accompanied by some change in the curves of saddle-nodes of limit cycles and homoclinic loops. There is no need to examine the entire sequence of events occuring in the bifurcation diagram in the (α, δ)-plane, because they do not lead to any new bifurcation curves or parameter regions. Following results of Buriev (1983), we only mention that increasing ϵ drives the biangle above the Andronov-Hopf curve N for $\gamma < \gamma_{\text{cr}} \approx 1.4$, and below this curve for $\gamma > \gamma_{\text{cr}}$.

The bifurcation diagram and the whole set of phase portraits of system (3.5.4) may now be considered as complete.

Phase portraits and their bifurcations. The results above suggest ten different regimes (corresponding to phase portraits) of dynamical behavior (see Fig. 3.5.4) of system (3.5.4) depending on the parameters. What do they have in common and how do they differ? What events occur when the parameters are changed?

First of all, we classify the phase portraits of the system by the number of attractors and their features:

Attractor:	Regions:
Single equilibrium	*1, 5*
Single limit cycle	*3, 8*
Two equilibria	*2*
Equilibrium and limit cycle	*4, 6, 7, 9, 10.*

Let us dwell at length on the structure of the different phase portraits. Phase portrait *1* is like many of those we considered earlier when analysing various predator-prey models. It has one globally attracting equilibrium and is of no particular interest. For the parameters from region *5* the system also has a unique globally stable equilibrium. The only distinction in the system's behavior consists in the character of the transitional processes of getting back to the equilibrium after the system has been perturbed. However, when one analyses concrete ecological situations this difference may be rather important.

In regions *3* and *8* the system reaches a globally attracting oscillatory regime. We encountered this regime earlier already. Qualitatively new is the behavior of system (3.5.4) for parameters from region *2*, where the phase portrait contains two notrivial stable equilibria. Depending on the initial condition, the system may end up in either of them. Note that any sufficiently strong perturbation of the system that increases or decreases the size of one of the populations while the other remains unchanged drives the system from, say, the equilibrium A_1 into the basin of attraction of A_2 (see Fig. 3.5.4, *2*).

The dynamical behavior is most diverse when both a stable equilibrium and a limit cycle exist simultaneously in the phase portrait. First of all, one should distinguish between the cases that the stable equilibrium lies outside (*4*) or inside the limit cycle (*6, 7, 9, 10*). In the first case the behavior of the system is similar to that in phase portrait *2*, with the difference that the stable equilibrium A_1 is replaced by a stable limit cycle around the equilibrium. Any strong enough perturbation of the oscillatory regime with respect to each of the components drives the system into the basin of attraction of the equilibrium A_2.

In parameter regions *6, 7, 9* and *10* we find abrupt excitation of oscillations (in the phase sense). If the system is in equilibrium any sufficiently strong perturbation drives it to the basin of attraction of oscillations around this equilibrium. In this case, the unstable limit cycle around this equilibrium is the boundary of the basin of attraction in regions *7, 9* and *10.* In region *6* the basin of attraction of the equilibrium is more complicated. It is bounded by the stable manifold of the saddle C, spiraling out of the unstable equilibrium A_1. Thus, that domain is something like a spiral winding around A_1.

This structure of the basin of attraction of A_2 leads to two interrelated effects. First, the neighborhood of the equilibrium A_1 is, in a way, "dynamically uncertain" (Bautin, Shil'nikov, 1980): however small a perturbation of the initial conditions in any direction may be, it can drive the system from the basin of attraction of the equilibrium into the basin of attraction of the limit cycle, and vice versa. Second, if the perturbation of that equilibrium A_2 grows gradually in a definite direction (for instance, along the straight line connecting A_1 and A_2 in the phase portrait) the system shows a qualitatively nonmonotunous response: it jumps between the two basins.

After this description of the phase portraits, we now proceed by describing the events that may occur in system (3.5.4) with respect to attractors (stable equilibria and stable limit cycles) if we gradually vary parameters.

When parameters change, stable equilibria may show five different types of behavior:

1. *Jump from one equilibrium to another.* For parameters from region *2* the system may be in one of the two stable equilibria A_1 or A_2. If the system is in equilibrium A_2, then changing parameters across the boundary $\mathcal{A}_1\mathcal{C}_1$ in the bifurcation diagram does not affect the behavior of the system. When parameters cross the curve $\mathcal{A}_1\mathcal{B}_1$ the equilibrium disappears: the stable node A_1 merges with the saddle C, and the system "jumps" to the equilibrium A_2. Similarly, if parameters are initially from region *2* and the system is in A_2 then it jumps to A_1 when the parameters move into region *1* by crossing the curve A_1B_2.
2. *Gradual excitation of oscillations.* When the parameters cross from region *1* into region *3* oscillations are gradually excited around the unique equilibrium. We have already run into this phenomenon, in particular, when studying systems (3.4.2) and (3.4.8). When the parameters cross from region *2* into region *4*, then there is a gradual excitation of oscillations on the condition that the system has initially been at the equilibrium A_1. Nothing happens if the system initially was at A_2.
3. *Abrupt excitation of oscillations.* When the parameters cross from regions *6* or *10* into region *3*, or from region *7* into region *8*, we encounter an abrupt excitation of oscillations. We have already run into this phenomenon, in particular, when studying system (3.4.3). In the mentioned regions, the initial values of the variables at the equilibrium are within the range of abrupt excitation of oscillations.
4. *Jump from an equilibrium to a distant limit cycle.* When the parameters cross from region *4* into region *3*, the system, if it is initially in the equilibrium A_2, moves to an oscillatory regime after the equilibrium disappears. The difference from the previous case is that the initial values of the variables at the equilibrium lie generally beyond the range that they have in the new stable oscillations.

5. *Appearance of oscillations from a saddle-node loop.* When the parametes cross from region *5* into region *3*, an excitation of oscillations with large amplitude takes place. The difference from the two previous cases is that the oscillations are of relaxation type, and most of the time the system remains in the neighborhood of what had been the equilibrium before the bifurcation.

Let us study the events that may occur if the system is in an oscillatory regime and the parameters are changed. There are four such events.

1. *Gradual decay of oscillations.* This phenomenon, reverse to the phenomenon of gradual excitation of oscillations, occurs for parameters crossing from region *3* into region *1*, as well as from region *4* into region *2*, provided that for parameters from region *4* the system was in the oscillatory regime.
2. *Abrupt termination of oscillations.* This phenomenon, reverse to the phenomenon of abrupt excitation of oscillations, occurs for parameters crossing from region *9* into region *5* and from region *10* into region *1*.
3. *Breaking up of oscillations in a homoclinic loop.* This occurs when the parameters cross from region *4* into region *5*, provided that for parameters from region *4* the system was in the oscillatory regime. As the parameters approach the bifurcation curve, the amplitude increases and the oscillations change to relaxation type.
4. *Termination of oscillations in a saddle-node bifurcation on a limit cycle.* This phenomenon is the reverse of the appearance of oscillations from a saddle-node loop. It occurs when the parameters cross from region *5* into region *3*. When the parameters approach the bifurcation curve, the oscillations retain their large amplitude and change to relaxation type.

This completes the list of bifurcations causing qualitative changes in system (3.5.4) under the variation of parameters.

Ecological interpretation. Let us discuss the ecological interpretation of the phase portraits for fixed parameters, as well as events occuring to stable equilibria and oscillations when the parameters are changed. Of the greatest interest are the situations when there is more than one attractor in the phase portrait of the system.

For parameters from region *2*, there are two stable equilibria A_1 and A_2 with basins of attraction separated by the stable manifold of the saddle C. Ecologically, A_1 is an equilibrium that is "inherited" from Volterra's model, and the predator and prey populations control each others' numbers. The equilibrium A_2 has a qualitatively different nature, which can easily be seen if we consider the limiting case $\epsilon \to 0$. Here, the equilibrium population of predator v_{A_2} tends to $v_{\max} = \frac{1}{\delta}(1/\alpha - \gamma)$. This expression is quite transparent: $1/\alpha$ is the maximum growth rate of the predator population in the absence of mortality and when prey is abundant; $(1/\alpha - \gamma)$ is the maximum reproduction rate of the predator population when the mortality is taken into account; $1/\delta$ is the reciprocal of the intensity of intraspecies competition among the predators for resources other than prey (proportional to the capacity of the ecological niche of the predator when prey is abundant). On the whole, $v_{\max}$ characterizes the capacity of the ecological niche of the predator and is similar to the quantity K in the logistic equation.

For $\epsilon \to 0$ the equilibrium size of the prey population u_{A_2} tends to $1/\epsilon$, that is, to the maximum density permitted by the resources available to the prey.

Thus, at A_2 the prey population is limited not by predation, but rather by the available resources, while the predator population is limited by the resources other than prey.

The behavior of the system in region *4* resembles that in region *2*. The only difference is that the predator and prey populations, while mutually controling each others' numbers, coexist not in a stationary, but in an oscillatory regime (given by a stable limit cycle around A_1).

The ecological interpretation of the behavior of the system when the equilibrium A_1 is perturbed is similar to that we gave when we studied system (3.4.8). The only difference is that the prey population, "having escaped" (in the population sense) from the predator, does not reproduce without bounds since its size stabilizes close to the value permitted by the level of available resources.

Let us proceed to the ecological interpretation of the events caused by changing parameters.

An examination of the bifurcation diagram of the system shows that a given qualitative change in the functioning of the ecosystem may be produced by changing different parameters. We restrict our analysis to events that occur when the parameter δ is changed, and the other parameters are fixed at given values. Note that δ describes the intensity of predator competition for resources other than prey, and $1/\delta$ is a quantity proportional to the capacity of the ecological niche of the predator when prey is abundant. We assume the parameter ϵ to be small ($\epsilon \ll 1$). Ecologically, this means that the density of the prey population, determined by the available resources in the absence of predation, is much larger than its density in the presence of predation.

When the parameter δ is changed, the equilibrium A_1 behaves as the unique equilibrium A in system (3.4.8). The only difference is that when the equilibrium disappears or the oscillations break up in the homoclinic loop the prey population in system (3.5.4) does not grow without bound (unlike for system (3.4.8)), but it stabilizes since the system moves over to A_2.

Consider now what happens to A_2 when the parameters are varied. Note that, as there exist a single equilibrium of the system for parameters from regions *1*, *3* and *10*, the difference between the equilibria A_1 (predator-prey interaction) and A_2 (interaction of either population with its resource) vanishes. We keep in mind that coexistence of predator and prey in system (3.4.8) (the particular case of (3.5.4) for $\epsilon = 0$) for the parameters lying above the curve of saddle-nodes is impossible. Therefore, we believe it is reasonable to assume, somewhat artificially, that a set of parameters located above and to the right of the biangle corresponds to the existence of an equilibrium of type A_2, whereas the region located below and to the left of the biangle corresponds to the existence of an equilibrium of type A_1. The artificiality of this assumption consists of the absence of a distinct boundary between these two sets of parameters. However, this division is convenient for our interpretations, which we do not claim to be absolutely rigorous.

When the capacity $(1/\delta)$ of the ecological niche of the predator changes, the behavior at the equilibrium A_2 depends qualitatively on the biotic potential of the predator when prey is abundant. This quantity, as mentioned above, is proportional to $1/\alpha$. We shall gradually increase the fixed value of the predator's biotic potential and keep track of the system. The entire range of α is divided into seven intervals (see Fig. 3.5.3c).

1. $\alpha > \alpha_{\mathcal{G}_2}$. For all these values of δ the system has only a single (stable) equilibrium of type A_2. The biotic potential of the predator is very small, therefore, the predator is unable to control the prey population. The equilibrium sizes of the predator and prey populations are determined by the available resources.
2. $\alpha_{\mathcal{G}_2} > \alpha > \alpha_{\mathcal{F}}$. This is a rather narrow interval where the equilibrium loses its stability gradually with a decrease in the capacity of the predator's ecological niche (boundary $\mathcal{F}\mathcal{G}_2$). Like every case of gradual excitation of oscillations, this phenomenon is reversible: when changing the parameter in the reverse direction, the equilibrium becomes stable for the same threshold of the parameter. This is an illustration of the statement that classification of the equilibria is rather artificial for parameters in the neighborhood of a "corner of the biangle".
3. $\alpha_{\mathcal{F}} > \alpha > \alpha_{\mathcal{I}_\infty}$. In this interval of the biotic potential the increase in the capacity of the predator's ecological niche leads to an abrupt excitation of oscillations, that is, oscillations characterized by large amplitudes.
   It is important that this effect, although the same all through the interval, may be brought about by different mechanisms: either the focus becomes unstable while the unstable limit cycle shrinks to it (curve $\mathcal{F}\mathcal{C}_2\mathcal{B}_2$ intersected) or the stable node and the saddle merge within the stable cycle (curve $\mathcal{B}_2\mathcal{I}_1$). This observation is worth stressing, because the criteria for approaching the dangerous parameter boundary are quite different in these two cases.
   The creation of large-amplitude oscillations here is locally and parametrically irrespective of the bifurcation mechanism: decreasing the capacity of the ecological niche below the value at which the oscillations appeared, does not lead to the disappearance of the oscillations. Only a further significant reduction in the capacity of the ecological niche (intersection of curve $\mathcal{F}\mathcal{J}\mathcal{I}_1$) results in an abrupt restoration of stationary coexistence. Again, the bifurcation mechanism of this restoration may be different: either a saddle-node of limit cycles (curve $\mathcal{F}\mathcal{J}\mathcal{H}$), or the death of the stable cycle in the homoclinic loop (curve $\mathcal{H}\mathcal{I}_1$).

4. $\alpha_{\mathcal{I}_\infty} > \alpha > \alpha_{\mathcal{D}_\infty}$. In this interval of the biotic potential the increased capacity of the ecological niche gives rise to an interesting phenomenon which has not been described before in terms of mathematical ecology. As the capacity increases above a certain threshold (curve $\mathcal{I}_1\mathcal{D}_1$), oscillations with large amplitude are generated, as in the previous case. The distinguishing feature of these oscillations is that they have a very large period (of the order $T \approx 1/\sqrt{\delta - \delta_{\text{cr}}}$). Besides, on the limit cycle most of the time is spent near the regime that was stable below threshold. Most interesting is that hysteresis is not observed here, unlike in the previous case of abrupt excitation of oscillations. When the parameter is decreased the equilibrium is restored at the same threshold.
   An ecological interpretation of this phenomenon in terms of stabilizing and destabilizing factors does not seem to be possible, because it involves all factors taken into account in the model.
5. $\alpha_{\mathcal{D}_\infty} > \alpha > \alpha_{\mathcal{C}_\infty}$. In this interval, the increased capacity of the ecological niche also leads to large-amplitude oscillations. However, unlike in the previous cases, the equilibrium values of the variables above $\alpha_{\mathcal{C}_\infty}$ lie far from the range of the variables during the oscillations in this interval. Therefore, it would be correct to speak of a jump from the equilibrium into oscillations in this case, rather than of an excitation of oscillations. As for $\alpha_{\mathcal{F}} > \alpha > \alpha_{\mathcal{I}_1}$, the phenomenon is parametrically and locally irreversible, that is, hysteresis occurs.
6. $\alpha_{\mathcal{C}_\infty} > \alpha > \alpha_{\mathcal{A}_\infty}$. The increased capacity of the ecological niche results in a change from the equilibrium A_2 to A_1. A decrease in the capacity does not immediately restore equilibrium A_2, hysteresis occurs. The capacity must drop a lot in order to restore the original equilibrium (curve $\mathcal{A}_1\mathcal{B}_1$).
   The ecological meaning of this effect is, to all appearance, as follows. At a small capacity of the niche, the predator population is not large enough to control the prey population, which is limited by its own resources (equilibrium A_2). As the capacity of the niche grows, the equilibrium density of the predator population increases up to a level at which the predator can perform the task of controlling the prey population. The population density of the prey (and, after that, of the predator) decreases, and the system jumps to A_1, at which the predator and prey populations control each other.
   After the equilibrium A_1 is established, decreasing the predator's resources other than prey decreases the predator population density, despite the fact that it is mainly governed by the interaction with the prey population. Correspondingly, the predator pressure on the prey population is weakened and the equilibrium density of the prey population increases. By further reducing the capacity of the ecological niche of the predator the equilibrium density of the prey population reaches so high a level that the predator is "unable" to control it. At that moment the population density of the prey (and after that, that of the predator) increases, and the system jumps back to A_2.

All events that occur by changing the capacity of the ecological niche of the predator for $\alpha_{\mathcal{D}_\infty} > \alpha > \alpha_{\mathcal{C}_\infty}$ can be interpreted in a totally analogous manner. The only difference is that the stationary coexistence of predator and prey is unstable when they mutually control their densities, and they can coexist only in an oscillatory regime corresponding to the stable limit cycle around A_1.

7. $\alpha < \alpha_{\mathcal{A}_\infty}$. Irrespective of the capacity of the ecological niche of the predator, the system has a unique equilibrium state. Ecologically, this implies that the biotic potential of the predator is so high that the predator has always control over the prey density. However, the stationary coexistence may be unstable, and in this case the populations coexist in an oscillatory regime.

This discussion completes the ecological interpretation of dynamical regimes and their rearrangements in system (3.5.4). Most interesting are the following results:

(a) it is possible to have stable coexistence of predator and prey in different states for the same external conditions, that is, for the same parameters,
(b) different hysteresis effects occur when the external conditions change, in other words, one finds irreversible qualitative rearranegements of regimes,
(c) non-hysteretic excitation of large-amplitude oscillations is possible.

Remark. Many important special dynamical features of multistable systems were studied by Harrison (1986), Wollkind *et. al.* (1988), Hainzl (1988), Collins, Wollkind (1990), Collins *et al.* (1990), Wolkind *et al.* (1991).

A model similar to system (3.5.4) was studied by A. S. Isaev and his colleagues (Isaev *et al.*, 1979*a*–*c*; Nedorezov, 1979). The common feature of these models is that they, unlike a large majority of other models of mathematical ecology, study a situation with more than one stable nontrivial equilibrium in the phase portrait. Isaev et al. used an approach going back to Kolmogorov (1936, 1972) by which only general limitations concerning the type of functions are made, whereas we analyse specific examples of the corresponding functions and how they depend on parameters. Kolmogorov's approach makes it possible to establish only some of the most general dynamical properties (the number of equilibria and, sometimes, their local stability). Concerning equally important related dynamical features, for instance the relative positions of stable and unstable manifolds and limit cycles, one must be satisfied with statements of the following type: A certain type of phase portrait "does not contradict the constraints imposed on the right-hand side of a system of ordinary differential equations" (Nedorezov, 1979), but does not necessarily occur.

On the other hand, the bifurcation diagram presented in this book, being structurally stable, retains its form when the right-hand sides of the differential equation is slightly changed. Therefore, it is valid not only for the systems studied here but also for all sufficiently close systems.

To describe the main dynamical characteristics of two interacting "phytophagan–entomophage" populations a five-parameter model has been suggested (Bazykin *et al.*, 1993; Bazykin, Berezovskaya, 1994, to appear). It is studied with the help of methods from bifurcation theory. There are good reasons to conjecture that the main dynamical events described by that model occur in the neighborhood of a codimension-four bifurcation point.

3.5.3. *Predator Saturation, Predator Competition for Prey and Competition among Prey*

Allowing for these factors leads to the system (Bazykin, Buriev, 1980, 1981)

$$\begin{aligned} \dot{x} &= ax - \frac{bxy}{(1+Ax)(1+By)} - ex^2 \\ \dot{y} &= -cy + \frac{dxy}{(1+Ax)(1+By)} . \end{aligned} \tag{3.5.7}$$

Setting $t = \tau/a$, $x = (a/d)u$, and $y = (a/b)v$ changes this system to

$$\begin{aligned} \dot{u} &= u - \frac{uv}{(1+\alpha u)(1+\beta v)} - \epsilon u^2 , \\ \dot{v} &= -\gamma v + \frac{uv}{(1+\alpha u)(1+\beta v)} , \end{aligned} \tag{3.5.8}$$

where $\alpha = aA/d$, $\beta = aB/b$, $\epsilon = ea/d$ and $\gamma = c/a$. Equation (3.5.8) depends on four parameters. To construct the complete bifurcation diagram we shall make use of the fact that for $\epsilon = 0$ system (3.5.8) is in fact (3.4.11), its bifurcation set is shown in Figs. 3.4.4*b* and 3.4.8*b*. Recall that the behavior of system (3.4.11) depends on the sign of $(\gamma - 1)$.

To study system (3.5.8), we first fix a value of $\gamma \leq 1$ and construct a one-parameter family of (α, β)-planes cutting through the three-dimensional parameter $(\alpha, \beta, \epsilon)$-space, where the parameter ϵ is successively increased from zero. Then we repeat this procedure for an arbitrary value of $\gamma > 1$.

Looking ahead, it should be noted that, depending on the parameters, the system has one, three or no nontrivial equilibria. The latter case occurs for $(\alpha + \epsilon > 1/\gamma)$ and will not be discussed further.

The bifurcation diagram and the phase portraits of system (3.5.8) for $\gamma \leq 1$ are equivalent to those of system (3.5.4). The only difference is that the parameter δ, which characterizes the intensity of predator competition for resources other than prey in system (3.5.4), has been replaced by the parameter β in (3.5.8), describing the intensity of predator competition for prey. Since the factors considered in models (3.5.4) and (3.5.7) are similar in their biological meaning, the interpretation of the results is in complete analogy.

The bifurcation diagram of system (3.5.8) for $\gamma > 1$ and $\epsilon = 0$ is equivalent to that of system (3.4.11). For $\gamma > 1$ and $\epsilon \ll 1$ the (α, β)-plane (cutting through the complete bifurcation diagram in $(\alpha, \beta, \epsilon)$-space) contains (like for $\gamma \leq 1$) a crescent-shaped biangular region of parameters for which there are three nontrivial equilibria in the phase portrait. The Andronov-Hopf curve for $\gamma > 1$ looks different in the (α, β)-plane than for $\gamma \leq 1$. Its location, relative to the coordinate axes and to the saddle-node bifurcation curves, can be determined if we compare the bifurcation diagram of system (3.5.8) for $\epsilon = 0$ and $\beta = 0$, again using continuity. The qualitative character of the location of three curves of homoclinic loops can be determined from our study of system (3.5.4). This was confirmed numerically by T.I. Buriev (Bazykin, Buriev, 1981).

It is important that the curve where the saddle quantity of the saddle is zero (dotted curve $\mathcal{B}_1\mathcal{B}_2$ in Fig. 3.5.6) intersects not only the "large" homoclinic loop but also the "small" one around the equilibrium A_1. The intersection point $\mathcal{H}_1$ is an end point for the curve of saddle-nodes of limit cycles surrounding A_1. The other end point of this curve lies on the the Andronov-Hopf curve of A_1. It marks the point $\mathcal{F}_1$ where the first Lyapunov quantity is zero. Thus, we have completed the construction of the bifurcation diagram in the (α, β)-plane for an arbitrary $\gamma > 1$ and $\epsilon \ll 1$.

Most parameter regions in Fig. 3.5.6 correspond to phase portraits which we have already described when studying system (3.5.4). (The numbering is the same.) Only for parameters from regions *11* and *12* the phase portraits of systems (3.5.8) and (3.5.4) differ (Fig. 3.5.7).

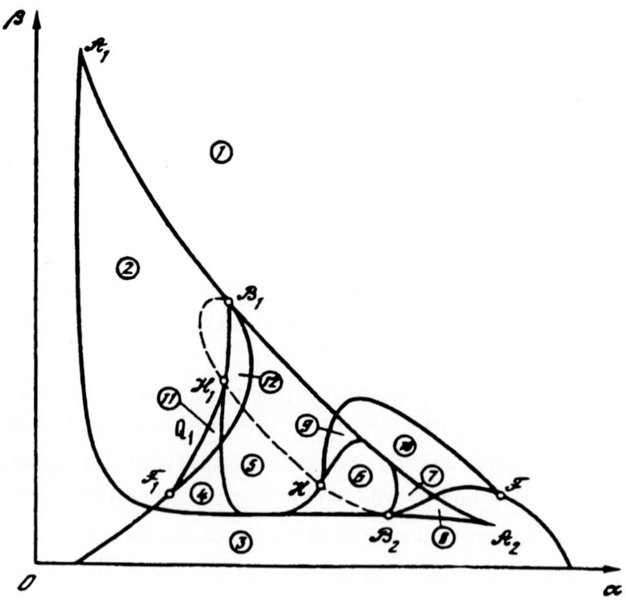

Fig. 3.5.6. The (α, β)-plane cutting through the bifurcation diagram of system (3.5.8) for $\epsilon \ll 1, \gamma > 1$.

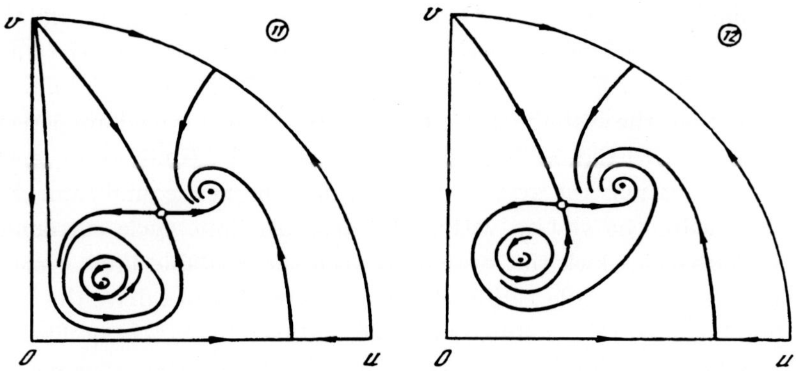

Fig. 3.5.7. Phase portraits of system (3.5.8) for parameters from regions *11* and *12*.

Now consider how the (α, β)-plane changes as ϵ is increased for fixed $\gamma > 1$. As ϵ grows, the bifurcation diagram of Fig. 3.5.6 successively undergoes two important changes related to the codimension-three bifurcations we have encountered before. The first change corresponds to a change of the direction in which the the Andronov-Hopf curve and the curve of saddle-nodes meet at the point $\mathcal{B}_1$. Furthermore, both end points $\mathcal{F}_1$ and $\mathcal{H}_1$ of the saddle-node of limit cycle curve Q shrink to the point $\mathcal{B}_1$ for the same value $\epsilon_{cr}(\gamma)$. We have met this codimension-three bifurcation when we studied system (3.4.16) (Fig. 3.4.15). As ϵ grows further, the bifurcation diagram in the (α, β)-plane takes the form in Fig. 3.5.3*c*, as regions *11* and *12* shrink to a point and disappear from the plane. No new parameter regions are formed.

Further increasing ϵ changes the (α, β)-plane qualitatively in the same way as the (α, δ)-plane of system (3.5.4). The main rearrangement in the (α, β)-plane occurs by the codimension-three bifurcation in which the Andronov-Hopf curve passes through the "corner" of the saddle-node curves. If the parameter ϵ is large enough, the biangle is either *above* or *below* the Andronov-Hopf curve, depending on the value of γ. When ϵ grows further, the biangle contracts to a point and disappears. Here again no new parameter regions and, correspondingly, no new phase portraits appear. The detailed qualitative investigation of system (3.5.8) is now complete.

Let us describe more closely phase portraits *11* and *12*, which have not been found earlier, as well as the rearrangements that occur when the parameters change. The behavior corresponding to phase portrait *12* is similar to that of the system in region *5*: two locally stable equilibria A_1 and A_2 are simultaneously present in the phase portrait. The only difference is the character of the boundary separating their basins of attraction. We have already mentioned that in region *5* the basin of attraction of A_1 is bounded by the stable manifold of the saddle, while in region *12*, the corresponding boundary is an unstable limit cycle surrounding this equilibrium.

For parameters from region *11* the system is more complicated. In this case (and for the first time in our study), the system has three nontrivial attractors at the same time: the stable equilibrium A_1, the stable limit cycle surrounding A_1, and the stable equilibrium A_2. In particular, this is evidence that perturbations of the equilibrium A_1 may lead to three different results, depending on their nature and intensity: for small perturbations the system returns to the initial state, for medium perturbations it shows transition to the oscillatory regime around the equilibrium, and finally, for strong perturbations the system is driven into the new equilibrium A_2. Small perturbations leave the system within the unstable limit cycle surrounding A_1. Medium perturbations kick the system between the unstable limit cycle and the stable manifold of the saddle C. Finally, strong perturbations drive the system out of the region bounded by the stable manifold and into the other equilibrium.

When studying system (3.5.4), we have described the events that may occur to stable equilibria and limit cycles by changing parameters. Is there anything new that can happen in system (3.5.8)?

1. *Disappearance of an equilibrium.* The equilibrium disappears when the parameters move from region *2* into region *1* by crossing the curve of stable saddle-nodes as in system (3.5.4). But this also happens when they move from region *12* into region *5* by crossing the Andronov-Hopf curve of the equilibrium A_1. In this case, A_1 loses its stability, and the equilibrium A_2 remains as the only attractor in the phase portrait. Phenomenologically, this event is similar to what occurs to A_1 by a transition from *2* to *1* when crossing the curve $\mathcal{A}_1\mathcal{B}_1$. But the mechanisms of these two phenomena and the criteria for approaching the boundary in parameter space are different.
2. *Abrupt excitation of oscillations.* They arise in system (3.5.8) around the equilibrium A_1, like those described for system (3.5.4), for a transitions from *10* into *3* (the only equilibrium A loses stability), as well as for the transitions from *6* to *3*, and *7* to *8* (A_2 becomes unstable). These oscillations occur when the parameters move from region *11* into region *4*. When the parameters cross from region *11* into region *12* the oscillations abruptly stop.

Summing up the results of the study of system (3.5.8), we cansay the following. Allowing for predator saturation and competition among prey, taking predator competition for prey into account results in a wider range of dynamical behavior than taking into account predator competition for resources other than prey. The main features of the dynamics, however, do not change drastically. The only new feature is the possibility of the simultaneous existence of three nontrivial attractors, namely two stable equilibria and a stable limit cycle around one of them, in a certain parameter region.

3.5.4. *Prey Competition and Competition among Predators for Resources Other than Prey (Type III Trophic Function)*

Allowing for these three factors gives the system

$$\dot{x} = ax - bx^2y/(1 + Ax^2) - ex^2 ,$$
$$\dot{y} = -cy + dx^2y/(1 + Ax^2) - hy^2 , \tag{3.5.9}$$

which after an appropriate change of variables takes the form

$$\dot{u} = u - u^2v/(1 + \alpha u^2) - \epsilon u^2 ,$$
$$\dot{v} = -\gamma v + u^2v/(1 + \alpha u^2) - \delta v^2 . \tag{3.5.10}$$

This model differs from system (3.5.2) because we are taking predator competition into account, and from system (3.5.4) because of the assumption of nonlinear predation at small prey population density.

A predator-prey system in which the nullclines are qualitatively as those considered in (3.5.10) was studied by Freedman (1979). He showed that the existence of two nontrivial equilibria is possible, one of which disappears when the parameters change.

The investigation of system (3.5.10) is an exact repitition of what we have done for system (3.5.8) and, therefore, it is not given here. The result is that the bifurcation diagram in $(\alpha, \delta, \epsilon, \gamma)$-space of system (3.5.10) and the bifurcation diagram in $(\alpha, \beta, \epsilon, \gamma)$ of system (3.5.8) are qualitatively the same. (Specific parameter values of bifurcations may naturally be different.) The biological interpretation of the dynamics and its changes depending on parameters is analogous to that in Sec. 3.5.3.

3.5.5. *Lower Critical Prey Density and Competition among Prey*

In Chapter 2 it was shown that allowing for both the nonlinear reproduction of an isolated population at small density and for natural mortality leads to a lower critical population density. Allowing for this effect for the prey population in a predator-prey system gives rise to a model which is, in a sense, biologically ill-defined (see Sec. 3.4.8). Within the framework of this model, both populations are doomed to die out for all initial non-zero population densities.

We take (2.2.9) as the equation describing the dynamics of prey population density in the absence of a predator, and keep the original second equation of Volterra's model for the predator population. This gives the system (Bazykin, Berezovskaya, 1979):

$$\dot{x} = ax(x - L)(K - x) - bxy ,$$
$$\dot{y} = -cy + dxy . \tag{3.5.11}$$

Setting $t = \tau/aK^2$, $x = Ku$ and $y = (aK^2/b)v$ changes this into

$$\dot{u} = u(u - l)(1 - u) - uv ,$$
$$\dot{v} = -\gamma v(m - u) , \tag{3.5.12}$$

where $l = L/K$, $\gamma = d/aK$, and $m = c/dK$.

The biological meaning of the parameters in this system is quite obvious: l is the ratio of the lower critical density of the prey population and the density determined by the prey's resources in the absence of a predator; γ is the coefficient of conversion of prey biomass into predator biomass (in scaled variables it may be less or bigger than one); m is a parameter that has different interpretations. First, m is the stationary population density of the prey population coexisting with the predator. It is natural to regard it as a measure of how well the predator is adapted to the prey: the lower the density of prey that can ensure a stationary existence of the predator, the better the predator is adapted to the prey. Second, m can be thought of as a measure of predator mortality (at a fixed value of γ), that is, again as a quantity that decreases when the preadator becomes increasingly adapted.

The equations of the nullclines of the system are

$$\begin{aligned} &\dot{u} = 0, \quad v = (u - l)(1 - u)\,; \\ &\dot{v} = 0, \quad u = m\,. \end{aligned} \tag{3.5.13}$$

Their possible relative positions are shown in Fig. 3.5.8a.

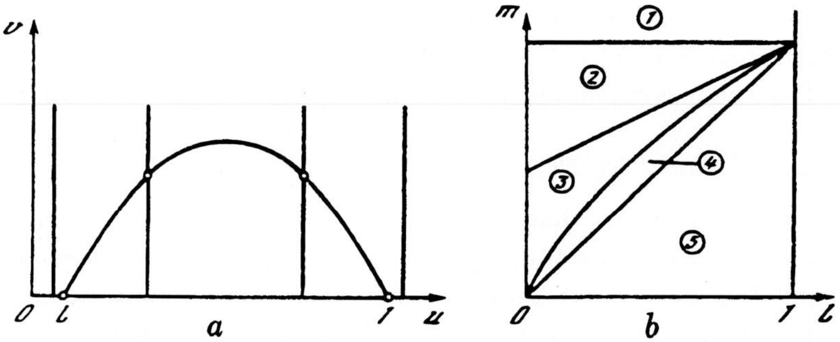

Fig. 3.5.8. Four possible relative positions of nullclines (a), and the bifurcation diagram (b) of system (3.5.12).

Note that the shapes and the locations of the nullclines, as well as the coordinates of equilibria and their stability, are independent of the value of γ. Therefore, as before it seems natural to construct the bifurcation diagram in the two-dimensional cross section of the complete bifurcation set given by the (l, m)-plane for a fixed value of γ. Again, we keep track of how this cross section changes when γ is changed. In this case, the biologically significant parameter region is the first octant of parameter space, given by $0 \leq l \leq 1$.

The location of some of the bifurcation curves in the (l, m)-plane is obvious without calculations. For all parameter values, the system has three equilibria O $(u = v = 0)$, $B_1(u = 1, v = 0)$ and $B_2(u = l, v = 0)$. The origin of the phase plane is always a stable node. Let us consider some of the events that occur in the phase portrait when we decrease m.

1. $m > 1$. B_2 is a saddle, and B_1 is a stable node. The stable manifold of B_1 separates the basins of attraction of O and B_1.
2. $1 > m\frac{1}{2}(l+1)$. B_1 and B_1 are saddles and A is either a stable node or focus. $m = 1$ is a bifurcation corresponding to the existence of a stable saddle-node AB_1 in the phase portrait.
3. $\frac{1}{2}(l+1) > m > l$. B_1 and B_1 are saddles and A is either an unstable focus or a node. $m = \frac{1}{2}(l+1)$ is a bifurcation at which A loses its stability. The mechanism of stability loss will be considered below.
4. $l > m > 0$. B_2 is an unstable node and B_1 is a saddle. The only attracting object is the origin O of the phase plane. For $m = l$ an unstable saddle-node AB_2 is formed when A retreats into the negative region.

We located three bifurcation curves in the bifurcation diagram, given by the conditions $m = l, m = \frac{1}{2}(l+1)$ and $m = l$. We now have a complete knowledge of the phase portraits for the parameters from regions *1* and *5*, as well as from *2* and *4*, in the vicinity of the bifurcation curves $m = 1$ and $m = l$. The structure of the phase portraits for parameters from a neighborhood on both sides of the Andronov-Hopf curve $m = \frac{1}{2}(l+1)$ is not clear yet.

It is obvious from the consideration of the phase portraits *2* and *4* (Fig. 3.5.9) that the bifurcation diagram in the triangle bounded by the ordinate and the bifurcation curves $m = 1$ and $m = l$ should contain at least one more bifurcation curve, namely a curve of heteroclinic connections where the unstable manifold of the saddle B_1 going into the interior of the first quadrant coincides with stable manifold of the saddle B_2.

Remark. For parameters from the mentioned triangle, the unstable manifold of B_2 coincides with stable manifold of B_1, because both of them are part of the abscissa of the phase portrait due to the specific form of the system. This implies that for the parameters, for which the unstable manifold of B_1 concides with stable manifold of B_2, there exists a heteroclinic cycle in the phase portrait, formed by the stable and unstable manifolds of the two saddles B_1 and B_1. Therefore, we call the corresponding bifurcation curve the heteroclinic cycle curve in the sequel. In a generic situation, the bifurcation corresponding to the formation of a heteroclinic cycle (or closed contour) of two saddles has codimension two (or more exactly, "one plus one"). However, owing to the specific form of system (3.5.12), one of the manifolds of each of the saddles B_1 and B_1 is formed by a coordinate axis, so that the above bifurcation has conditional codimension one and leads to a curve in the (l, m)-plane.

Apart from the curve of homoclinic loops, the bifurcation set may in general contain curves of saddle-nodes of limit cycles. None of them has an analytical parametrization, and they must be found numerically. The numerical construction of the bifurcation diagram becomes a lot easier if we use the following analytical arguments:

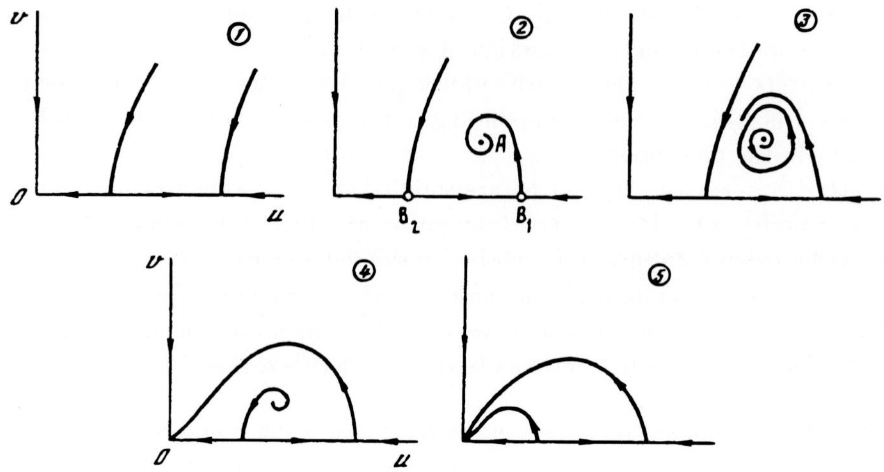

Fig. 3.5.9. Structurally stable phase portraits of system (3.5.12).

1. The first Lyapunov quantity is positive everywhere on the Andronov-Hopf curve $m = \frac{1}{2}(l+1)$ of the equilibrium A: if we vary the parameters across the Andronov-Hopf curve from above then A becomes unstable and gives rise to a small stable limit cycle in the phase portrait.
2. A study by F.S. Berezovskaya (Bazykin, Berezovskaya, 1979) into the structure of equilibria at infinity, and into complex equilibria in the finite part of the phase portraits (that appear for $l = 0$, $l = 1$, $m = 0$ and $m = 1$) revealed the following. First, the curve of heteroclinic cycles in the bifurcation diagram starts at the point $m = l = 1$ below the Andronov-Hopf curve A and ends at the point $m = l = 0$. That means that the heteroclinic cycle curve crosses the Andronov-Hopf curve of A either not at all, or it does so an even number of times. Second, the curve of saddle-nodes of limit cycles, if it exists, has no end points on the curves $m = l$ and $m = 1$. This implies that the curve of saddle-nodes of limit cycles, if it exists and is not closed, must have its end points on the curve of heteroclinic cycles. Those end points, if present in the bifurcation diagram, correspond to the codimension-two bifurcation of a neutral heteroclinic cycle.
3. The neutrality of the heteroclinic cycle is given by a simple local condition, namely by the ratio of eigenvalues of the system at the saddles B_1 and B_1. The heteroclinic cycle is stable provided that

$$|\lambda^-_{B_1}\lambda^-_{B_2}/\lambda^+_{B_1}\lambda^+_{B_2}| > 1\,, \tag{3.5.14}$$

where the upper indices corresponds to the sign of an eigenvalue. The heteroclinic cycle is unstable if the above expression is less than one and neutral if it is equal to one. For system (3.5.12), the heteroclinic cycle is neutral on the line $m = 2l/(1+l)$. In the bifurcation diagram, this line connects the origin of the (l, m)-plane with the point $m = l = 1$, and it lies below the Andronov-Hopf curve of A.

We have now made all possible analytical considerations concerning the structure of the bifurcation diagram of system (3.5.12). It is a comparatively simple numerical procedure to determine the relative positions of the stable and unstable manifolds of the saddles B_1 and B_2 for parameters satisfying the above neutrality condition. In this fashion it can easily be checked that that the heteroclinic cycle is always stable within a wide range of values of γ. The bifurcation diagram of system (3.5.12) is now complete (Fig. 3.5.8*b*).

Let us proceed with the biological interpretation of the results we obtained. The model predicts four different regimes of dynamical behavior. We list them in order of increasing predator adaptation to prey, that is, in order of decreasing m. All other parameters of the system may be assumed to be fixed.

1. For small predator adaption to prey ($m > 1$) the predator population always dies out. In other words, the density of the prey population, limited by the available resources, is insufficient to sustain the predator. Depending on the initial values, the prey population may either stabilize or also die out.
2. Increasing the predator's adaptation can result in a stable stationary coexistence of predator and prey.
3. Increasing the predator's adaptation further, the coexistence is possible only in an oscillatory regime; the amplitude of the oscillations increases with the predator's adaptation.
4. Finally, when the predator is very well adapted to the prey (For $m < m_{\rm sc}$, where $m_{\rm sc}$ is a value at which the heteroclinic cycle is formed), the predator becomes "overadapted", that is, it would be completely content with a prey population whose density were much smaller than the current one. Both populations are now doomed to die out for all initial values.

We should stress one important thing. The structure of the bifurcation diagram in the (m, l)-plane, in particular the relative position of the Andronov-Hopf curve and the curve of heteroclinic cycles, as well as the absence of a curve of saddle-nodes of limit cycles, was determined for the specific form of the function describing the prey dynamics in the absence of a predator. However, the main results, like the extinction of the predator for low adaptation, the coexistence of predator and prey for intermediate adaptation, and the extinction of both populations for excessive adaptation of the predator, remain valid for the much wider class of models of the form

$$\begin{aligned} \dot{x} &= ax(x-l)(1-x)f(x) - bxy\,, \\ \dot{y} &= -cy + dxy\,, \end{aligned} \qquad (3.5.15)$$

where $f(x)$ is a positive function with biologically natural conditions on continuity and smoothness. In other words, the main results of the study of system (3.5.12) remain valid if the prey population is limited by external resources in the absence of a predator and has a lower special density. Different modifications in parameter and phase space of the mentioned model were examined by Brauer and Soudack (1979).

In summary, the most interesting dynamics in the study of three-parameter modifications of the Volterra model (when allowing for competition among prey) can be found in those models that exhibit multistability and oscillatory regimes in different combinations.

Within the framework of our model, we have analysed the structure of a neighborhood in parameter space of a degenerate point of codimension three, near which all the mentioned behavior does occur. The phenomena we described are structurally stable, andtherefore our description of parameter space is valid not only for this specific model but also for all sufficiently close systems.

Appendix

A 3.1
Phase portrait of Eq. 3.1.2

$$\dot{x} = x - xy$$
$$\dot{y} = -\gamma y + xy$$

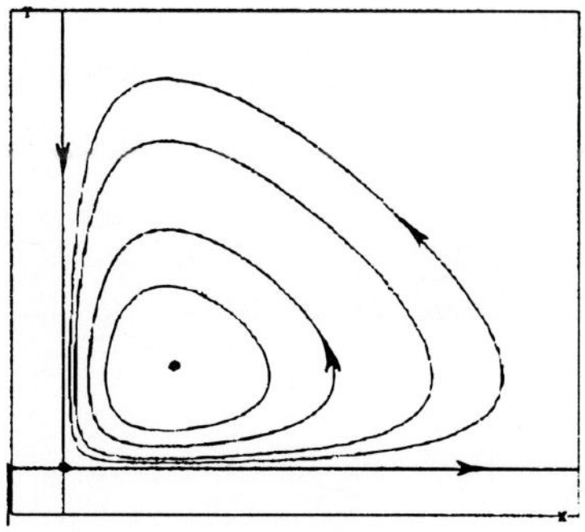

$\gamma=1$

A 3.3.1
Phase portrait of Eq. 3.3.2

$$\dot{x} = \frac{x^2}{1+x} - xy$$
$$\dot{y} = -\gamma y + \kappa xy$$

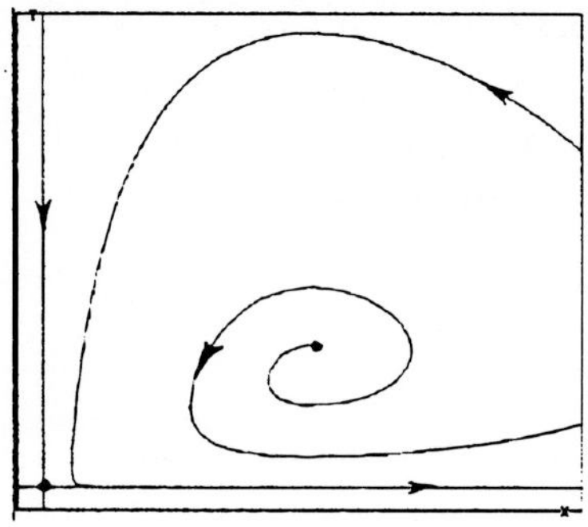

$\gamma=1$
$\kappa=1$

Phase portraits of Eq. 3.3.4

$$\dot{x} = x(1-x) - xy$$
$$\dot{y} = -\gamma y + \kappa xy$$

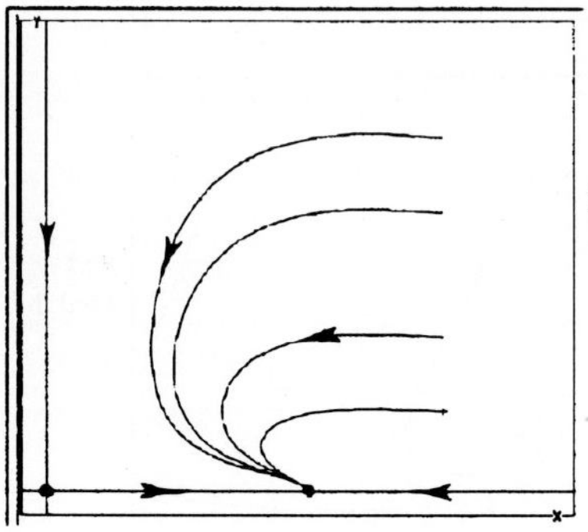

$\gamma=1$
$\kappa=0.$

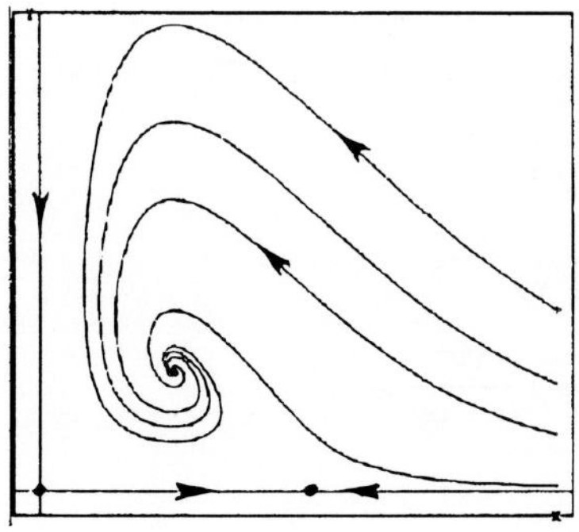

$\gamma/\kappa<1$

$\gamma=1$
$\kappa=2$

A 3.3.2
Phase portrait
of Eq. 3.3.6

$$\dot{x} = x - \frac{xy}{1+\alpha x}$$

$$\dot{y} = -\gamma y + \frac{xy}{1+\alpha x}$$

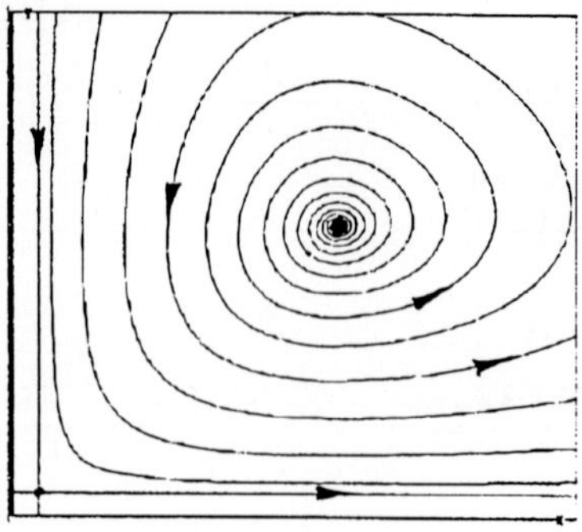

A 3.3.3
Phase portrait
of Eq. 3.3.8

$$\dot{x} = x - \frac{x^2 y}{1+\alpha x}$$

$$\dot{y} = -\gamma y + \frac{x^2 y}{1+\alpha x}$$

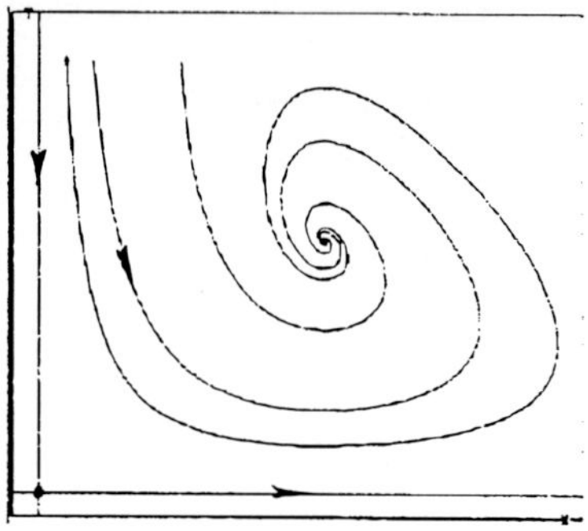

A 3.3.4
Phase portrait
of Eq. 3.3.10

$$\dot{x} = x - \frac{xy}{1+\beta y}$$

$$\dot{y} = -\gamma y + \frac{xy}{1+\beta y}$$

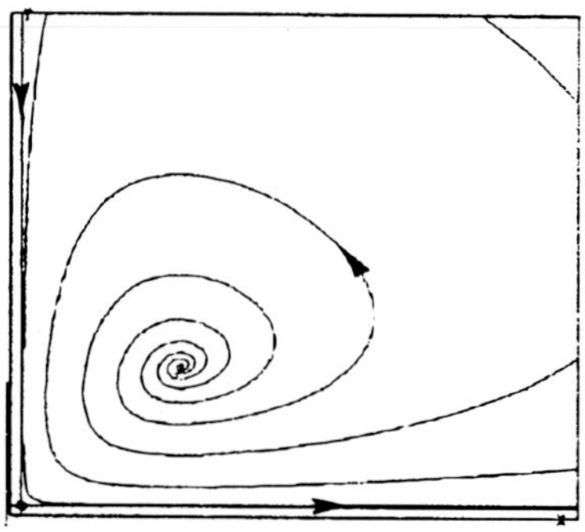

A 3.3.5
Phase portrait
of Eq. 3.3.14

$$\dot{x} = x - xy$$

$$\dot{y} = -\gamma y + \frac{xy^2}{1+\nu y}$$

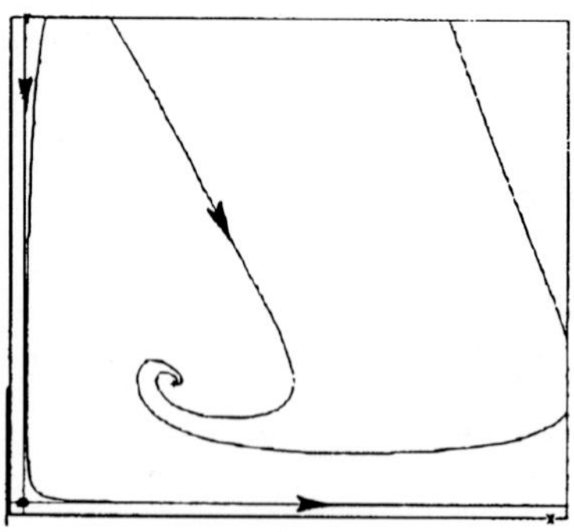

A 3.4.1
Phase portraits
of Eq. 3.4.2

$$\dot{x} = x - \frac{xy}{1+\alpha x} - \varepsilon x^2$$

$$\dot{y} = -\gamma y[1 - \frac{x}{1+\alpha x}]$$

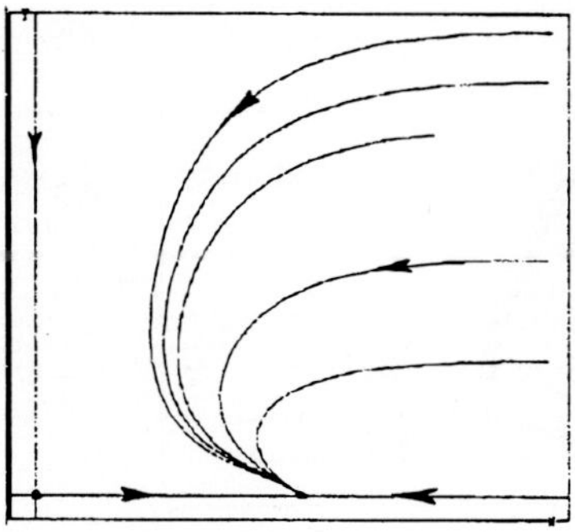

Region 1

α=0.5
γ=1
ε=1

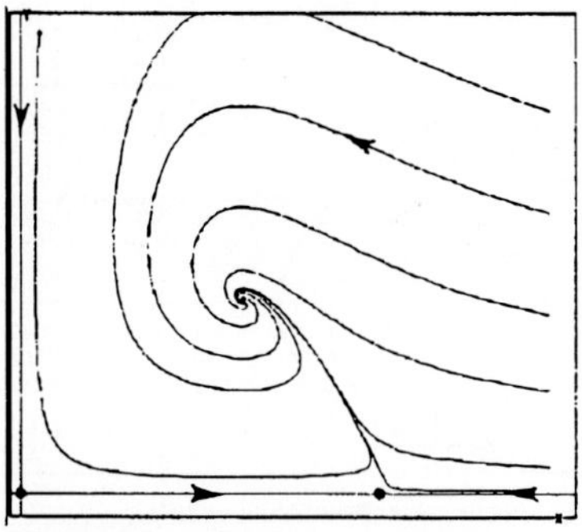

Region 2

α=0.5
γ=1
ε=0.3

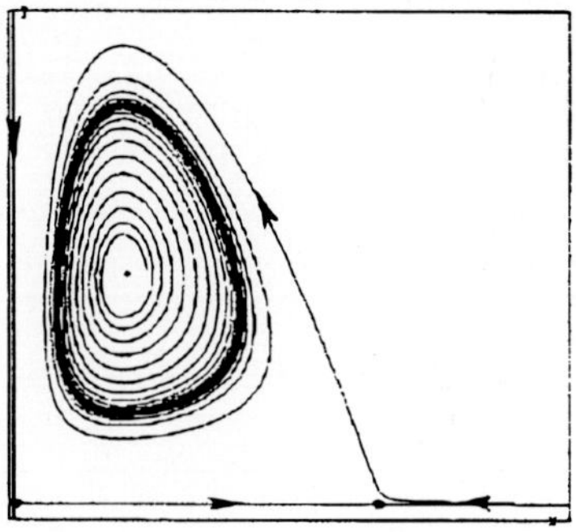

Region 3

α=0.5
γ=1
ε=0.15

A 3.4.2
Phase portraits
of Eq. 3.4.4

$$\dot{x} = \frac{x^2(1-x)}{n+x} - xy$$

$$\dot{y} = -\gamma y(m-x)$$

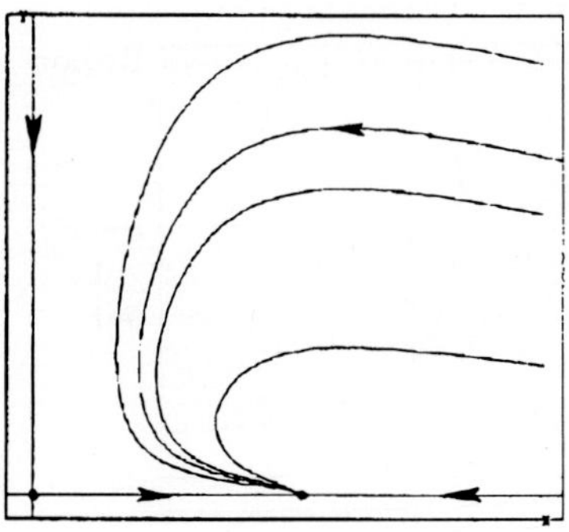

Region 1

n=0.05
γ=1
m=1.2

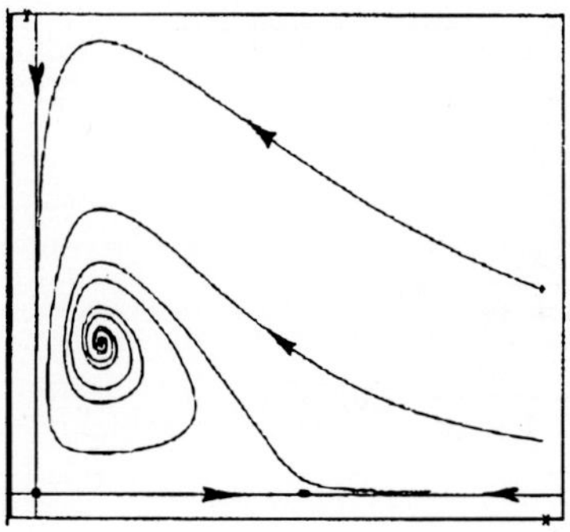

Region 2

n=0.05
γ=1
m=0.25

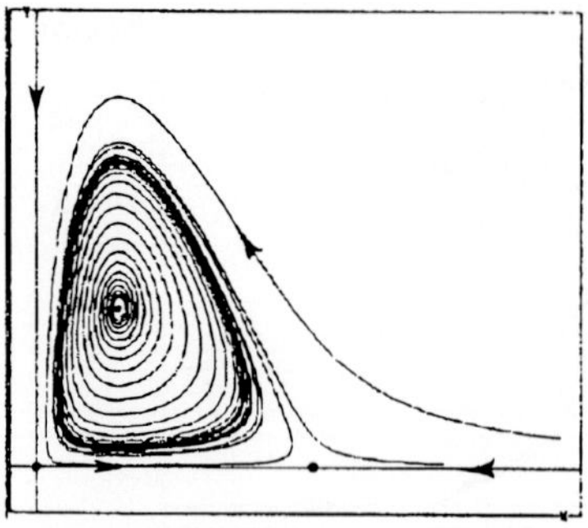

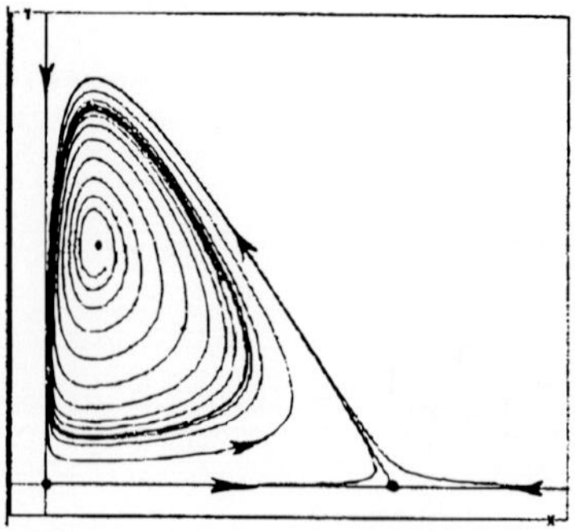

A 3.4.3
Phase portraits of Eq. 3.4.6

$$\dot{x} = x - \frac{x^2 y}{1+\alpha x^2}$$

$$\dot{y} = -\gamma y + \frac{x^2 y}{1+\alpha x^2}$$

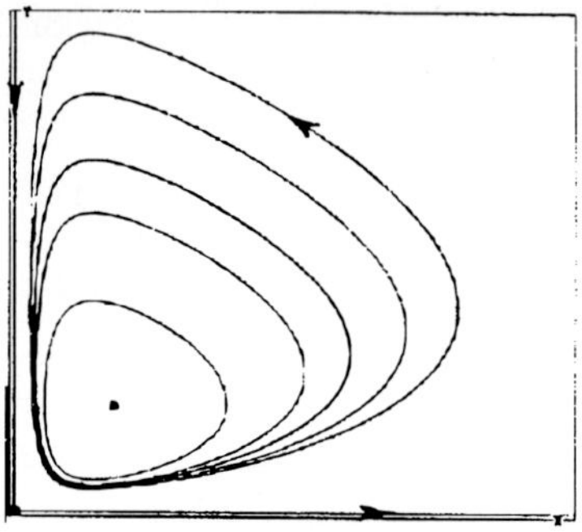

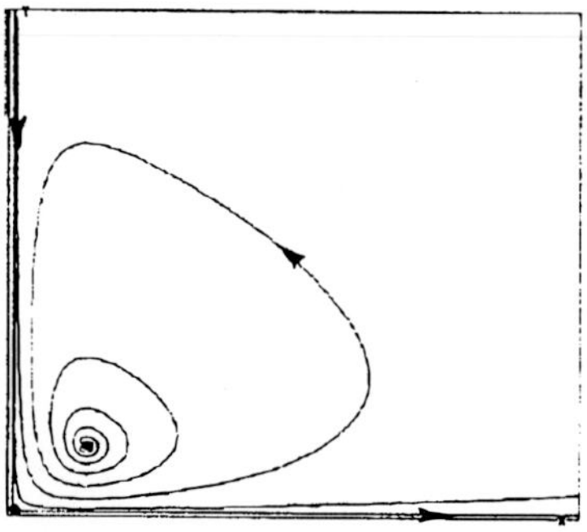

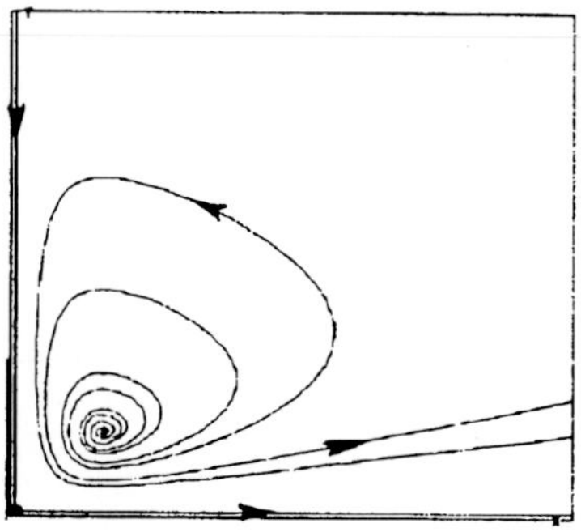

A 3.4.4
Phase portraits of Eq. 3.4.8

$$\dot{x} = x - \frac{xy}{1+\alpha x} - xy$$

$$\dot{y} = -\gamma y + \frac{xy}{1+\alpha x} - \delta y^2$$

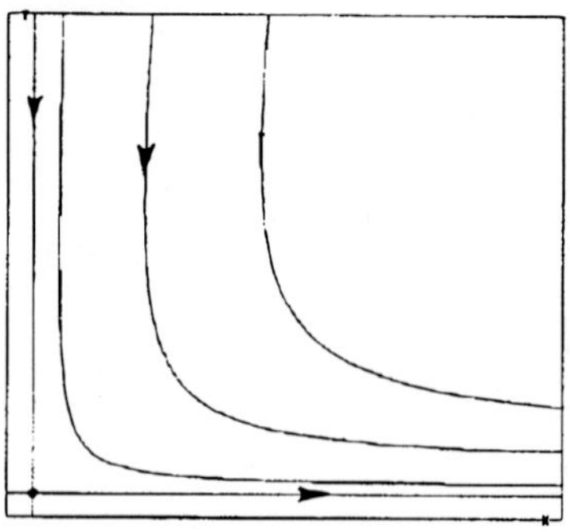

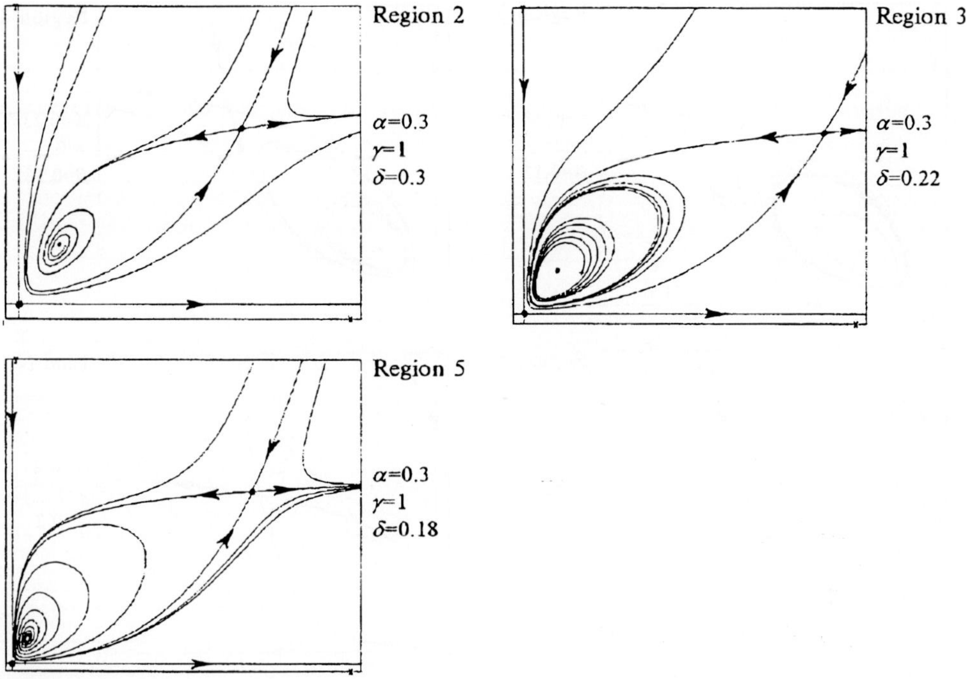

A 3.4.5
Phase portraits
of Eq. 3.4.11

$$\dot{x} = x - \frac{xy}{(1+\alpha x)(1+\beta y)}$$

$$\dot{y} = -\gamma y + \frac{xy}{(1+\alpha x)(1+\beta y)}$$

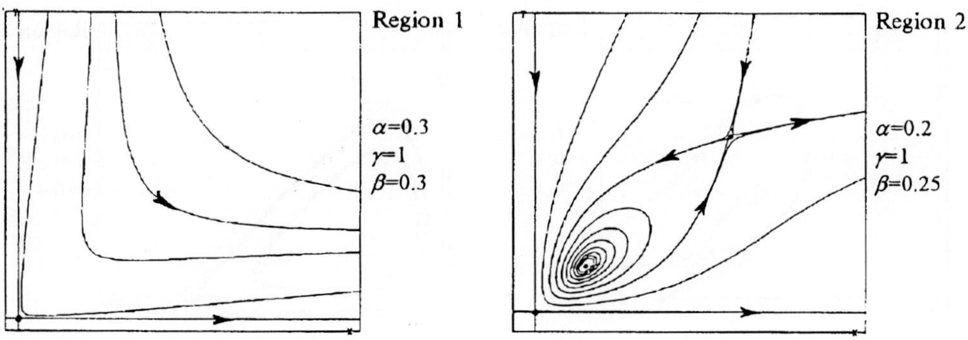

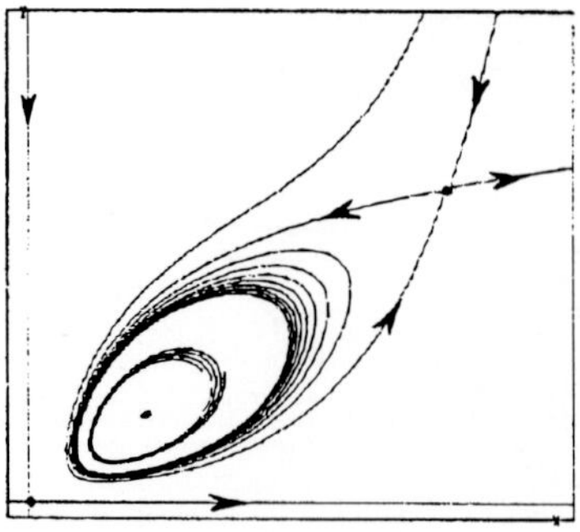

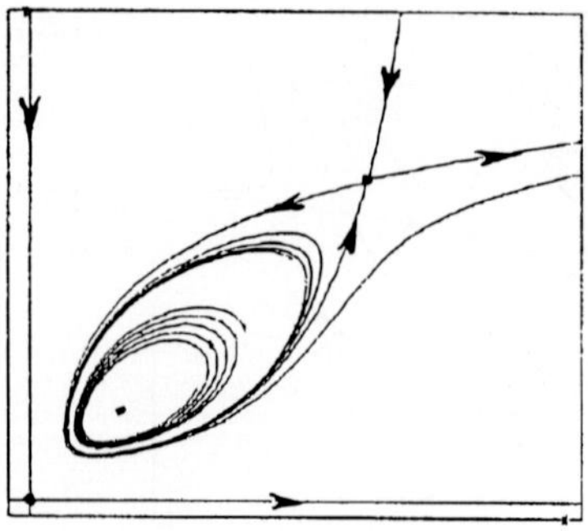

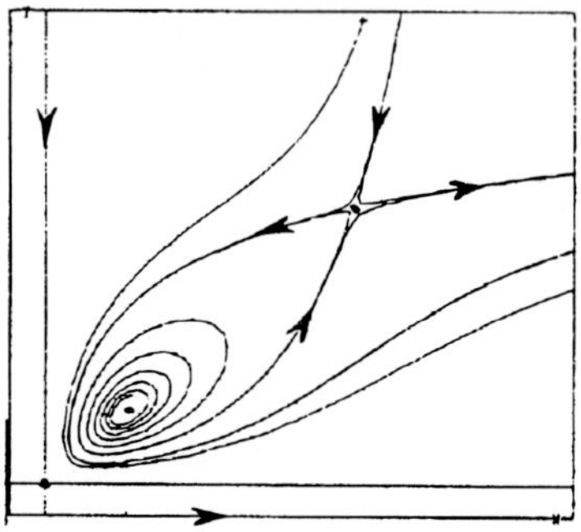

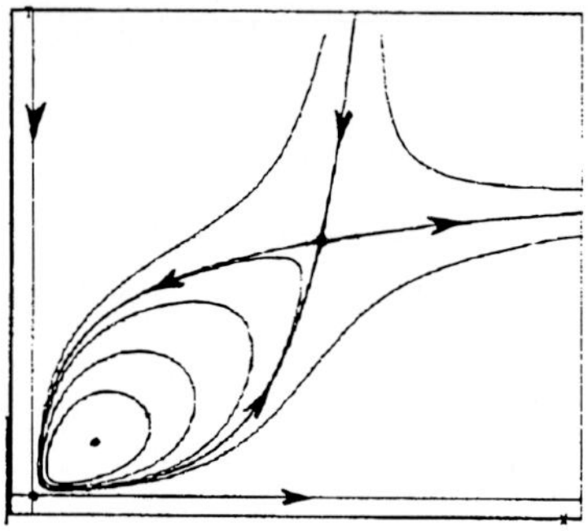

A 3.4.6
Phase portraits
of Eq. 3.4.16

$$\dot{x} = x - xy - \varepsilon x^2$$

$$\dot{y} = -\gamma y + \frac{xy^2}{n+y}$$

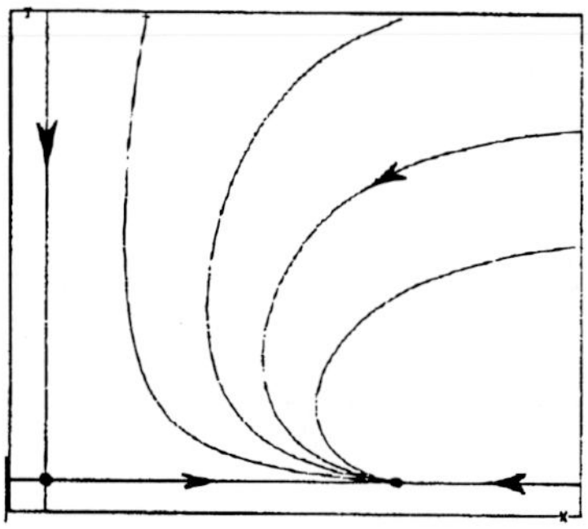

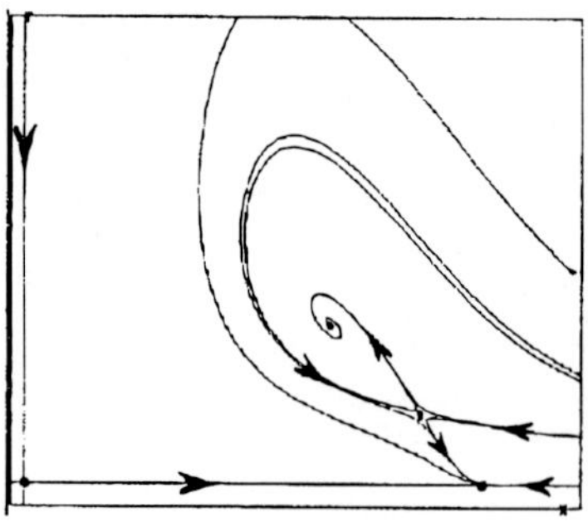

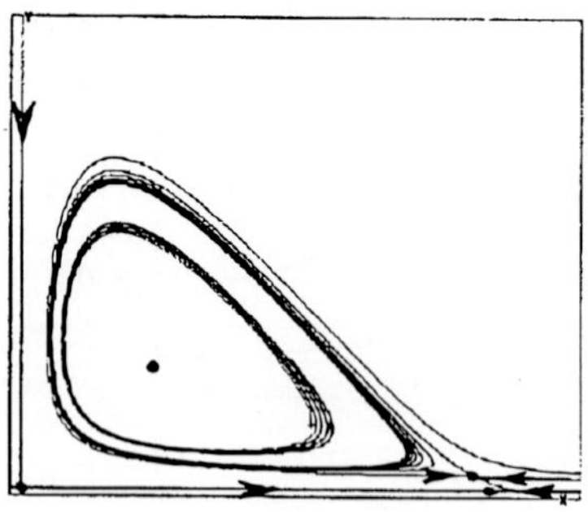

A 3.4.8
Phase portraits
of Eq. 3.4.21

$$\dot{x} = \frac{x^2}{1+x} - \xi x - xy$$
$$\dot{y} = -\gamma y + \kappa xy$$

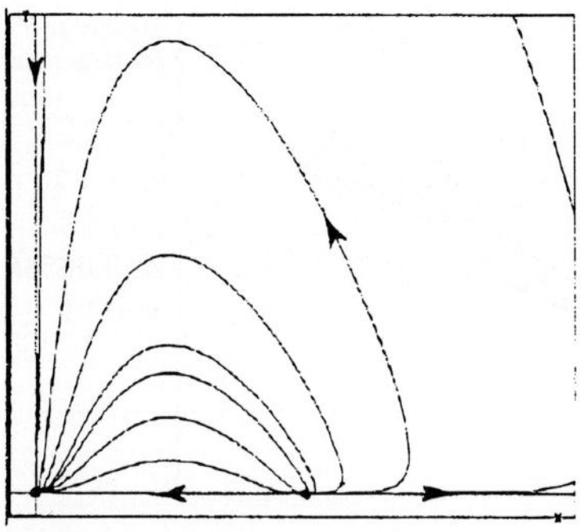

$\gamma/\kappa < \xi/(1-\xi)$

$\gamma=1$
$\xi=0.5$
$\kappa=2$

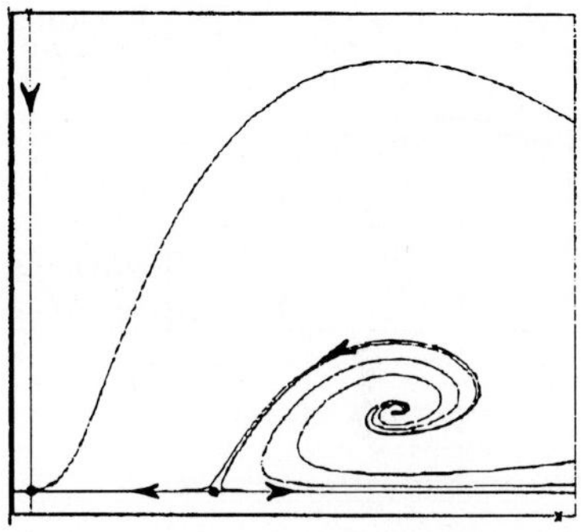

$\gamma/\kappa < \xi/(1-\xi)$

$\gamma=1$
$\xi=0.5$
$\kappa=0.5$

A 3.5.2
Phase portraits
of Eq. 3.5.4

$$\dot{x} = x - \frac{xy}{1+\alpha x} - \varepsilon x^2$$
$$\dot{y} = -\gamma y + \frac{xy}{1+\alpha x} - \delta y^2$$

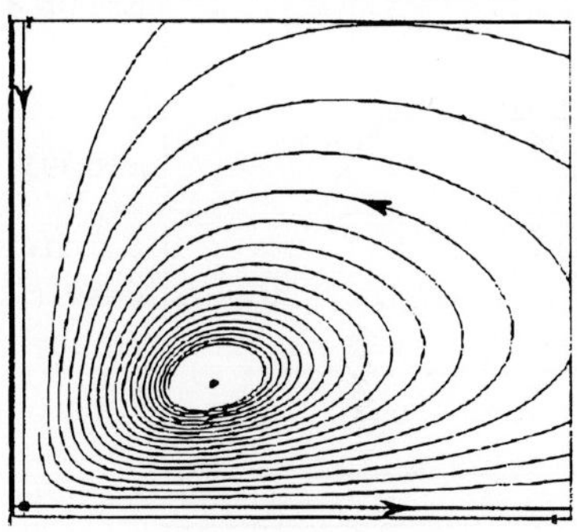

Region 1

$\alpha=0.2$
$\gamma=1$
$\delta=0.2$
$\varepsilon=0.01$

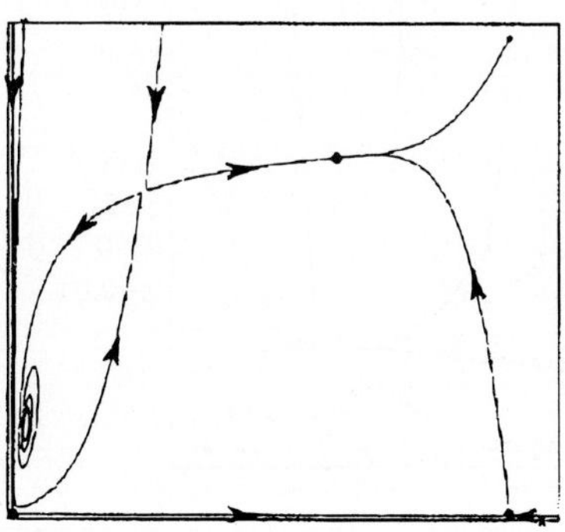

Region 2

$\alpha=0.3$
$\gamma=1$
$\delta=0.3$
$\varepsilon=0.01$

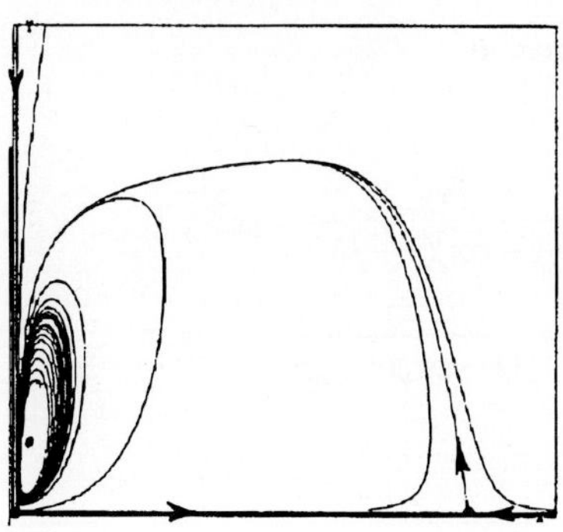

Region 3

$\alpha=0.3$
$\gamma=1$
$\delta=0.2$
$\varepsilon=0.01$

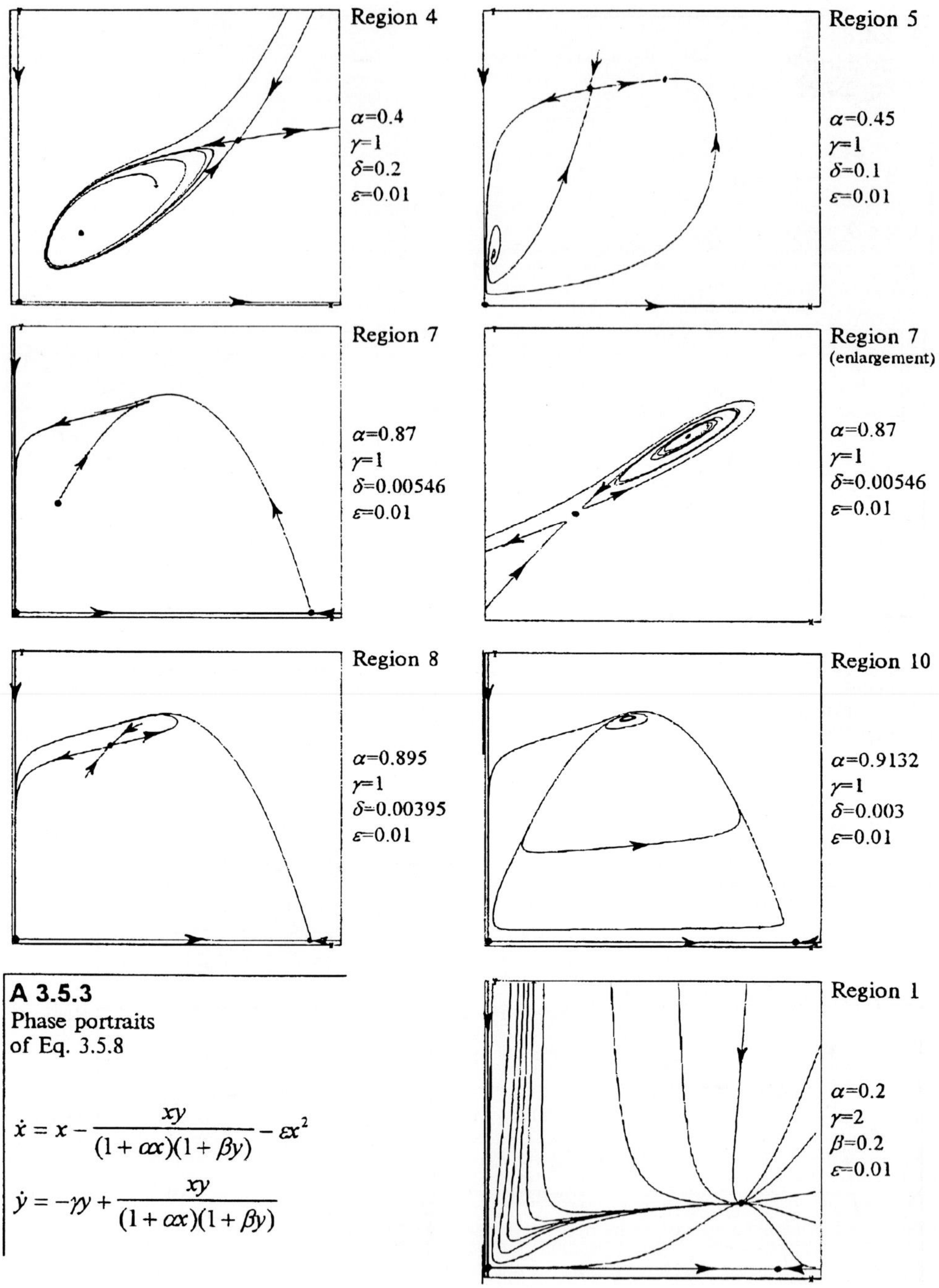

A 3.5.3
Phase portraits of Eq. 3.5.8

$$\dot{x} = x - \frac{xy}{(1+\alpha x)(1+\beta y)} - \varepsilon x^2$$

$$\dot{y} = -\gamma y + \frac{xy}{(1+\alpha x)(1+\beta y)}$$

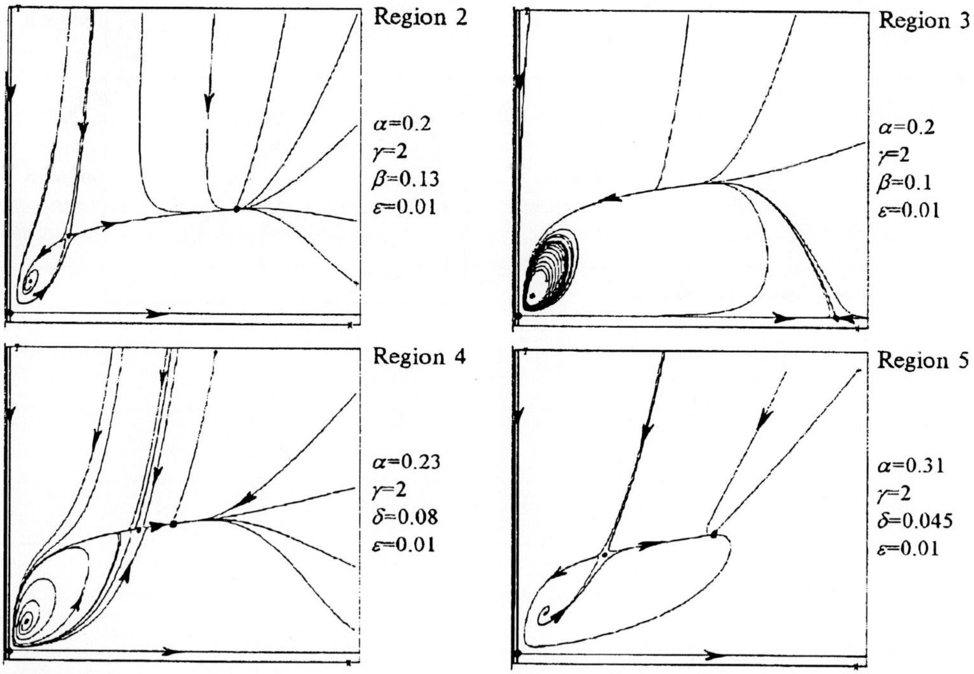

A 3.5.5
Phase portraits of Eq. 3.5.12

$$\dot{x} = x(x - l)(1 - x) - xy$$
$$\dot{y} = -\gamma y(m - x)$$

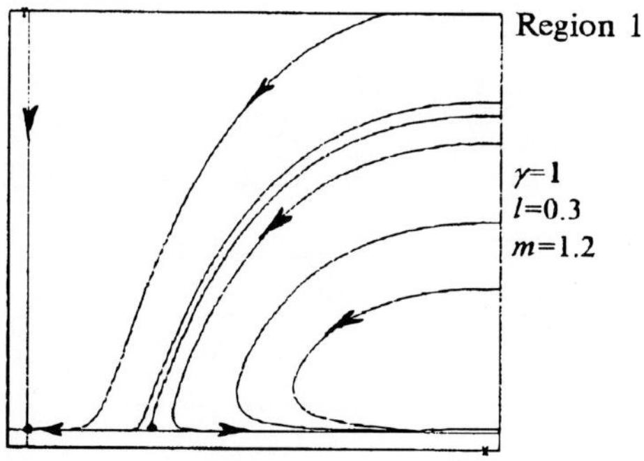

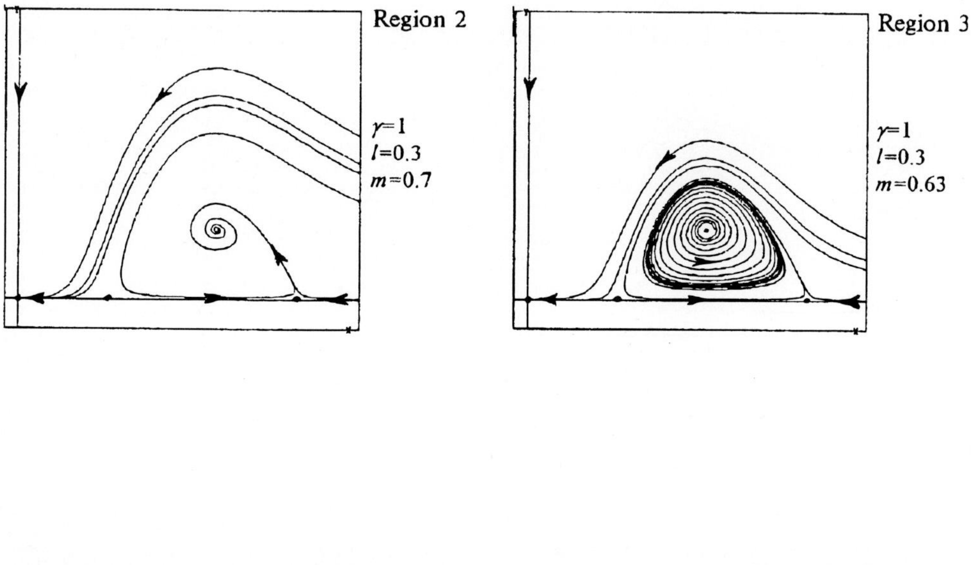

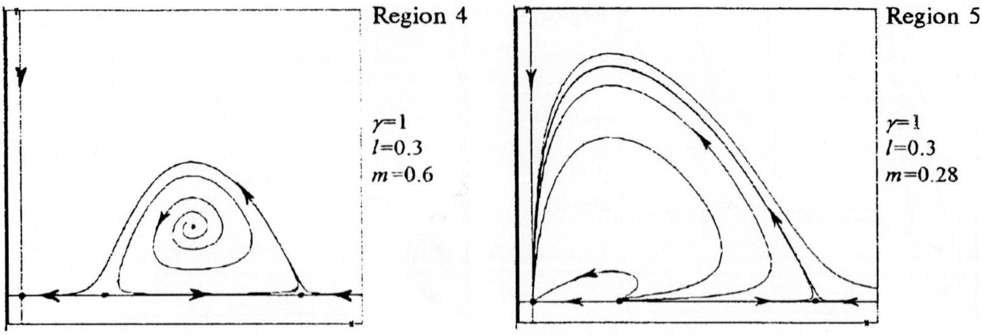
Region 4
γ=1
l=0.3
m=0.6
Region 5
γ=1
l=0.3
m=0.28

Chapter 4

Competition and Symbiosis

As we mentioned in Chapter 3, predator–prey interaction is the most promising one of the three main types of interspecies interactions (competition, symbiosis, and predator–prey interaction) in terms of the diversity of the possible dynamics. In this chapter, we consider competitive and symbiotic interactions. Their dynamics is considerably less rich, but also of great importance for the functioning of ecosystems.

4.1. Competition

4.1.1. *Two Logistic Populations*

To describe the dynamics of two competing populations, Volterra (1926), Lotka (1925) and then Gause (1933) proposed the set of equations

$$\begin{aligned} \dot{x}_1 &= a_1 x_1 (K_1 - x_1 - \alpha_2 x_2)/K_1\,, \\ \dot{x}_2 &= a_2 x_2 (K_2 - x_2 - \alpha_1 x_1)/K_2\,, \end{aligned} \tag{4.1.1}$$

which is a natural generalization of the logistic equation. Here a_1 and a_2 are coefficients of exponential growth of both populations at small densities (that is, in the absence of intra- and interspecies competition), K_1 and K_2 are the capacities of the ecological niches for both populations, and α_1 and α_2 are the coefficients of interspecies competition. This system can be rewritten in the equivalent form

$$\begin{aligned} \dot{x}_1 &= a_1 x_1 - e_{11} x_1^2 - e_{12} x_1 x_2\,, \\ \dot{x}_2 &= a_2 x_2 - e_{22} x_2^2 - e_{21} x_2 x_1\,, \end{aligned} \tag{4.1.2}$$

where e_{ij} are the coefficients of intra- and interspecies competition. Setting $t = \tau/a_1$, $x_1 = (a_1/e_{11})u_1$ and $x_2 = (a_2/e_{22})u_2$ changes system (4.1.2) to

$$\begin{aligned} \dot{u}_1 &= u_1 (1 - u_1 - \epsilon_1 u_2)\,, \\ \dot{u}_2 &= \gamma u_2 (1 - u_2 - \epsilon_2 u_1)\,. \end{aligned} \tag{4.1.3}$$

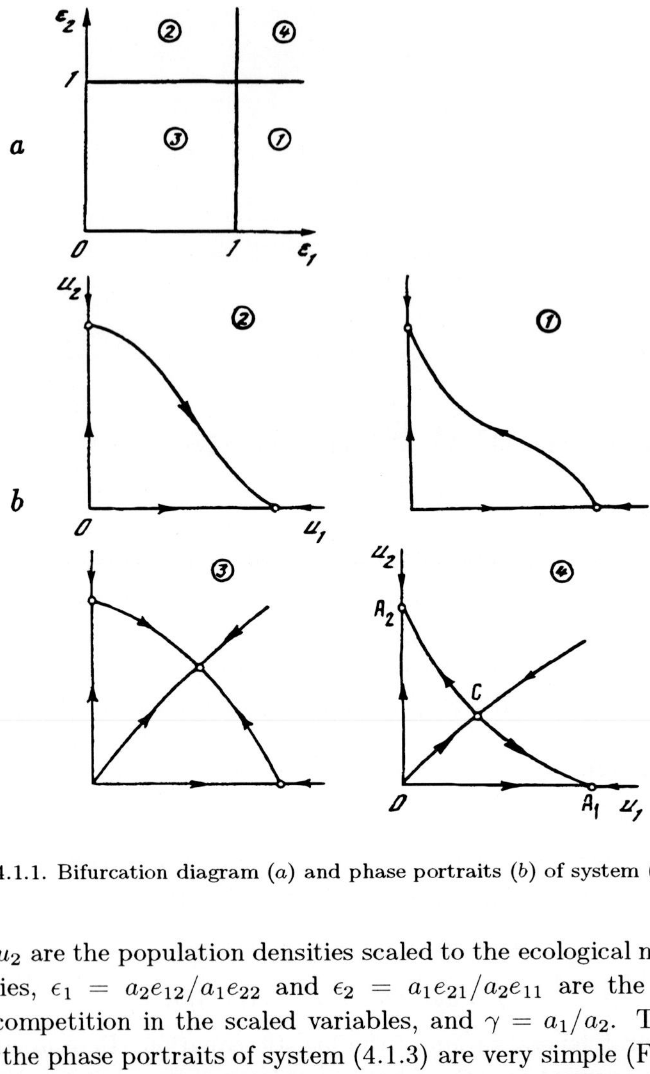

Fig. 4.1.1. Bifurcation diagram (a) and phase portraits (b) of system (4.1.3).

Here u_1 and u_2 are the population densities scaled to the ecological niche capacities of both species, $\epsilon_1 = a_2 e_{12}/a_1 e_{22}$ and $\epsilon_2 = a_1 e_{21}/a_2 e_{11}$ are the coefficients of interspecies competition in the scaled variables, and $\gamma = a_1/a_2$. The bifurcation diagram and the phase portraits of system (4.1.3) are very simple (Fig. 4.1.1). The behavior of the system is independent of γ. For $\epsilon_1 > 1$ and $\epsilon_2 < 1$ the second species is outcompeted by the first and dies out (Fig. 4.1.1b, region *1*). For $\epsilon_1 < 1$ and $\epsilon_2 > 1$ the situation is the opposite and the first species dies out (Fig. 4.1.1b, region *2*).

For $\max\{\epsilon_1, \epsilon_2\} < 1$ the interspecies competition for each of the species is less strong than the intraspecies competition, and competing species can coexist (Fig. 4.1.1b, region *3*). The limit situation here corresponds to $\epsilon_1 = \epsilon_2 = 0$ when the interspecies competition disappears and the species "do not pay attention to each other". However, when $\min\{\epsilon_1, \epsilon_2\} > 1$ the interspecies competition for each of the species is stronger than the intraspecies competition. In this case, a stable coexistence of the species is not possible: depending on the initial condition, one of them always dies out (Fig. 4.1.1b, region *4*). The basins of attraction of the equilibria A_1 and A_2, corresponding to the death of one of the rivals, are separated by the stable manifold of the saddle C, which conrresponds to an unstable coexistence of both populations.

The bifurcations that take place in the system when the parameters change have a natural biological interpretation. Most interesting here is the bifurcation that occurs when the parameters are initially in region *4*. Without loss of generality, we suppose that the system is at the equilibrium $A_2\{u_1 = 0,\ u_2 = 1\}$. Assume that the parameter ϵ_1 remains unchanged and that ϵ_2 gradually decreases. This decrease does not affect the equilibrium and, therefore, cannot be noticed by an external observer. In this case, if there is a weak invasion of individuals of the second species into the system, their equilibrium density remains at a certain low level. However, a further decrease in ϵ_2 narrows the basin of attraction of A_2, bounded by the abscissa and the unstable manifold of the saddle C. It finally shrinks to the unit interval and disappears when ϵ_2 reaches the threshold $\epsilon_2 = 1$. The equilibrium $A_1\{u_1 = 1,\ u_2 = 0\}$, corresponding to the death of the second species, becomes globally stable. In case of an invasion of the second species, however weak it may be, an external observer does notice that the equilibrium is perturbed. Then suddenly, the second species is ousted by the first.

4.1.2. *One of the Populations has a Lower Threshold Size*

The equations describing the population dynamics in this case have the form

$$\begin{aligned} \dot{x}_1 &= a_1 x_1 (x_1 - L)(K - x_1) - e_{12} x_1 x_2 , \\ \dot{x}_2 &= a_2 x_2 - e_{22} x_2^2 - e_{21} x_1 x_2 . \end{aligned} \tag{4.1.4}$$

Setting $t = \tau / a_1 K^2$, $x_1 = K u_1$ and $x_2 = (a_2/e_{22}) u_2$, this system becomes

$$\begin{aligned} \dot{u}_1 &= u_1 (u_1 - 1)(1 - u_1) - \epsilon_1 u_1 u_2 , \\ \dot{u}_2 &= \gamma u_2 (1 - u_2 - \epsilon_2 u_1) . \end{aligned} \tag{4.1.5}$$

Here $l = L/K$ is the lower threshold size of the first population, $\epsilon_1 = a_2 e_{12} / a_1 e_{22} K^2$ and $\epsilon_2 = e_{21} K / a_2$ are the coefficients of interspecies competition, and $\gamma = a_2 / a_1 K^2$.

The equations of the (nontrivial) nullclines of the system are

$$\begin{aligned} \dot{u}_1 = 0 :, &\quad u_2 = (1/\epsilon_1)(u_1 - l)(1 - u_1) ; \\ \dot{u}_2 = 0 :, &\quad u_2 = 1 - \epsilon_2 u_1 . \end{aligned}$$

The four possible relative positions of the nullclines are illustrated in Fig. 4.1.2*a*. This results in the bifurcation diagram in the (ϵ_1, ϵ_2)-plane of Fig. 4.1.2*b*, with the corresponding phase portraits for each region in Fig. 4.1.3. The phase portraits are independent of γ and are completely determined by the relative positions of the nullclines.

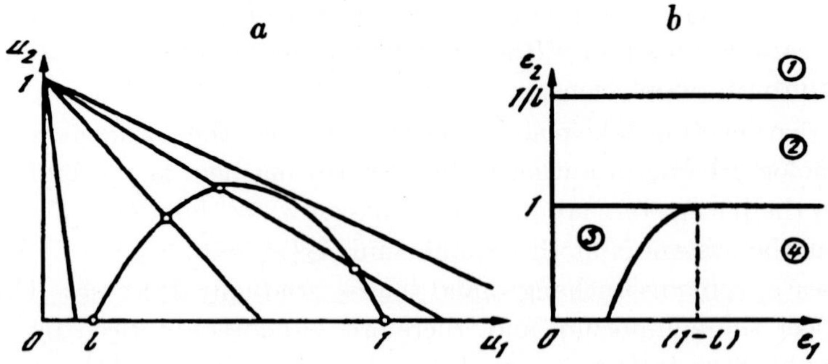

Fig. 4.1.2. Possible relative positions of nullcline (a) and the bifurcation diagram (b) of system (4.1.5).

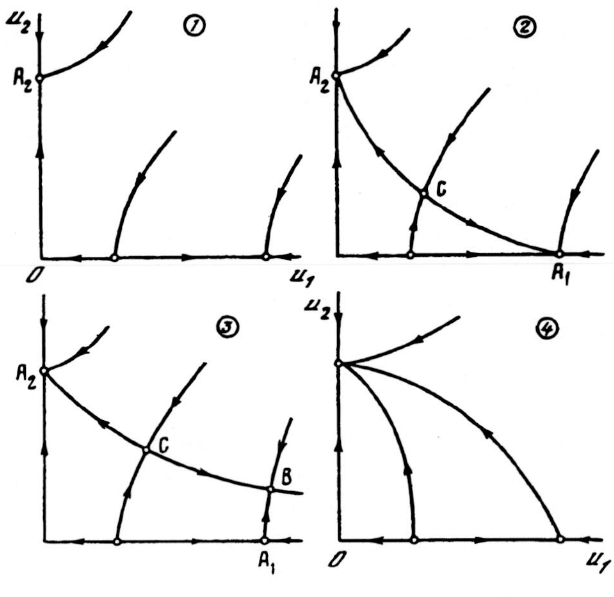

Fig. 4.1.3. Phase portraits of system (4.1.5).

The bifurcation diagram and the phase portraits of the system have a natural biological interpretation. There can be three stable equilibria: A_2 corresponds to the stable existence of the second species (the one without a threshold size) in the absence of the first. A_1 corresponds to the stable existence of the first species in the absence of the second. Finally, B corresponds to the stable coexistence of both populations. The equilibrium A_2 exists and is locally stable for all values of the parameters, whereas A_1 and B exist and are locally stable only for certain values of the parameters. That the existence of only the second species is always locally stable is understandable: immigration of small numbers of indivuduals of the first species, not exceeding its lower threshold size, cannot survive in the ecosystem.

The main distinction from the case of two competing logistic populations studied above is that there is a region of parameters (region *3*) where two regimes, the sole existence of the second population and the coexistence of the populations, are both locally stable. Assume that initially the parameters lie in region *3*, the value of ϵ_1 is not too small ($\epsilon_1 > (1-l)^2/4$) and the system is at B (stable coexistence of both populations). Let us now see what happens when ϵ_2, describing the competitiveness of the first species relative to the second, changes gradually. As ϵ_2 grows, the negative effect of the first species on the second gets more pronounced. The equilibrium size of the second species gradually drops to zero, so that the second species becomes extinct. The equilibrium density of the first species grows gradually until it reaches the limit value $u_1 = 1$, determined by the capacity of the ecological niche.

As ϵ_2 decreases gradually, the events develop in qualitatively different way. As long as ϵ_2 is above the curve separating region *3* and region *4*, decreasing this parameter, as expected, results in a gradual decrease of the size of the first and an increase of the size of the second species. This continues until ϵ_2 reaches a threshold at the boundary of regions *3* and *4*. It is important that the equilibrium B (stable coexistence of the populations) gradually approaches the boundary of its basin of attraction. This means that the situation is about to become dangerous, in spite of the fact that the total area of the basin of attraction of the equilibrium actually increases. Finally, imperceptible for an external observer (when ϵ_2 reaches the boundary between regions *3* and *4*) B reaches the boundary of its basin of attraction and becomes semi-stable. In other words, this is a saddle-node bifurcation of the saddle-node BC. By decreasing ϵ_2 further the previously stable situation is distroyed: the first species dies out, and the second reaches the equilibrium size $u_2 = 1$, determined by the capacity of its ecological niche. It should be stressed once again that at the moment preceding the extinction of the first species, its equilibrium size (when it coexists with the second species) is considerable: it is much more than a half of the equilibrium size determined by the capacity of its ecological niche in the absence of the second species.

4.1.3. *Two Populations with Lower Threshold Sizes*

The equations describing the population dynamics in this case can be written in the form

$$\begin{aligned} \dot{x}_1 &= a_1 x_1 (x_1 - L_1)(K_1 - x_1) - e_1 x_1 x_2 \,, \\ \dot{x}_2 &= a_2 x_2 (x_2 - L_2)(K_2 - x_2) - e_2 x_1 x_2 \,. \end{aligned} \tag{4.1.6}$$

Setting $t = \tau / a_1 K_1^2$, $x_1 = K_1 u_1$ and $x_2 = K_2 u_2$ gives

$$\begin{aligned} \dot{u}_1 &= u_1 (u_1 - l_1)(1 - u_1) - \epsilon_1 u_1 u_2 \,, \\ \dot{u}_2 &= \gamma u_2 (u_2 - l_2)(1 - u_2) - \epsilon_2 u_1] \,, \end{aligned} \tag{4.1.7}$$

where $\gamma = a_2 K_2^2 / a_1 K_1^2$, $\epsilon_1 = e_1 K_2 / a_1 K_1^2$, and $\epsilon_2 = e_2 K_1 / a_2 K_2^2$. The equations of the nullclines of the system are

$$\begin{aligned} \dot{u}_1 = 0, \quad u_2 &= \frac{1}{\epsilon_1}(u_1 - l_1)(1 - u_1)\,; \\ \dot{u}_2 = 0, \quad u_1 &= \frac{1}{\epsilon_2}(u_2 - l_2)(1 - u_2)\,. \end{aligned} \tag{4.1.8}$$

There are three possible relative positions of the nullclines (Fig. 4.1.4*a*), which correspond to three regions of the bifurcation diagram in the (ϵ_1, ϵ_2)-plane (Fig. 4.1.4*b*). Again, the relative positions of the nullclines determine the phase portraits completely (Fig. 4.1.5). The qualitative character of the bifurcation diagram in the (ϵ_1, ϵ_2)-plane (in fact a cross section of the complete four-dimensional bifurcation diagram) is independent of the parameters l_1 and l_2, as long as the latter are between zero and one. (As in earlier cases, the parameter γ does not affect the number and stability of equilibria and can be ignored.)

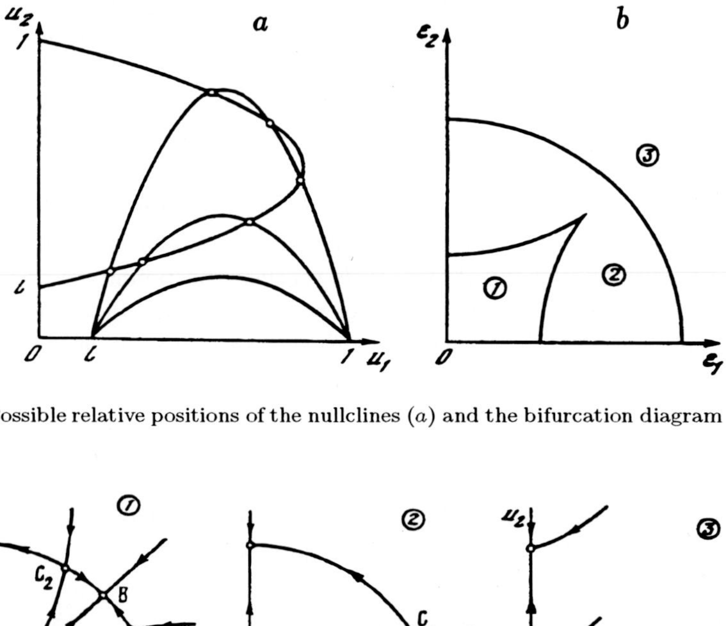

Fig. 4.1.4. Possible relative positions of the nullclines (a) and the bifurcation diagram (b) of system (4.1.7).

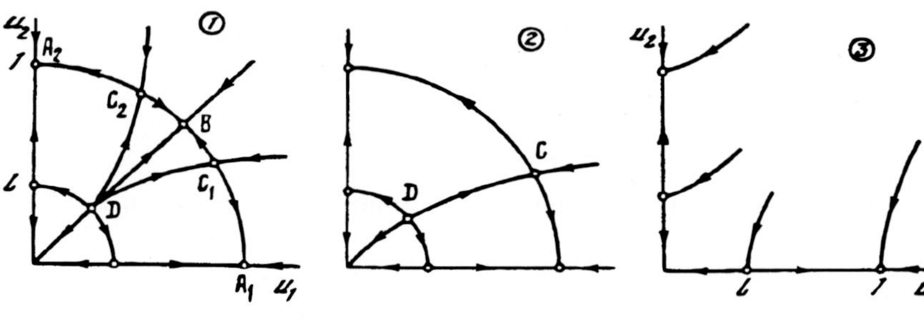

Fig. 4.1.5. Phase portraits of system (4.1.7).

It can be seen from the phase portraits that the trivial equilibrium O (both populations absent), as well as the equilibria A_1 and A_1 (existence of one population in the absence of the other) are locally stable for all values of the parameters. This is biologically obvious: it follows from the fact that both populations have lower threshold sizes. Furthermore, in region *1* the equilibrium B of coexisting populations is also locally stable (together with the above-mentioned equilibria O, A_1 and A_2). Its basin of attraction is bounded by the stable manifolds of the saddles C_1 and C_2, which both come from the unstable node D.

The events occurring to B when the parameters are varied have a natural biological interpretation. As the competitiveness of a species decreases, its equilibrium size also starts decreasing gradually, while the size of the competing species increases. It should be noted that these variations may turn out to be very small, and, therefore, difficult to observe. When the competitiveness is further decreased the corresponding parameter reaches a threshold at which the stable equilibrium B (corresponding to the coexistence of the populations) hits the boundary of its basin of attraction. Then a saddle-node bifurcation occurs, B disappears, and the less competitive species suddenly dies out.

Thus, two types of attractors are possible in the community of two competing populations:

(a) a stationary coexistence of populations, and
(b) existence od only one of the populations, withstanding a possible invasion by the other.

These regimes may be both globally and locally stable. When each of the populations has a lower threshold size, the mentioned regimes are locally stable. A variation in the life conditions of the populations may lead to a sudden perturbance of either of the equilibria.

4.2. Symbiosis

It is accepted to distinguish between two types of symbiotic relationships (Odum, 1971):

1. *protocooperation*: interspecies interaction is useful for both species, but is not necessary: each population may exist without a partner, and
2. *mutualism*: the interspecies interaction is necessary as a condition for the existence of each of the species. Each of them dies out in the absence of the partner.

Let us consider these interactions.

4.2.1. *Protocooperation*

Assume that in the absence of a partner the dynamics of both protocooperating populations can be described by a logistic equation with coefficients different for both populations in general. Regarding the effect of protocooperation itself, it is natural to describe it in first approximation as a predator–prey interaction, that is, by means of bilinear terms that have positive signs in both equations. The corresponding system of differential equations is

$$\dot{x}_1 = a_1 x_1 \frac{K_1 - x_1}{K_1} + P_1 x_1 x_2 \,,$$
$$\dot{x}_2 = a_2 x_2 \frac{K_2 - x_2}{K_2} + P_2 x_1 x_2 \,. \tag{4.2.1}$$

Setting $t = \tau / a_1$, $x_1 = K_1 u_1$ and $x_2 = K_2 u_2$ changes the system to

$$\dot{u}_1 = u_1(1 - u_1 + p_1 u_2) \,,$$
$$\dot{u}_2 = \gamma u_2(1 - u_2 + p_2 u_1) \,, \tag{4.2.2}$$

where $\gamma = a_2/a_1$, $p_1 = P_1 K_1 K_2 / a_1$ and $p_2 = P_2 K_1 K_2 / a_2$.

Figure 4.2.1 shows two possible relative positions of the nullclines of the system, together with the corresponding phase portraits.

A natural interpretation of these findings is the following. For $p_1 p_2 < 1$ the system has only one stable equilibrium corresponding to a stable coexistence of both populations. In this case, the equilibrium sizes of both populations are larger than those in the absence of the partner. This circumstance makes the situation different from the coexistence under competition, and it emphasizes the positive role of protocooperation for both populations.

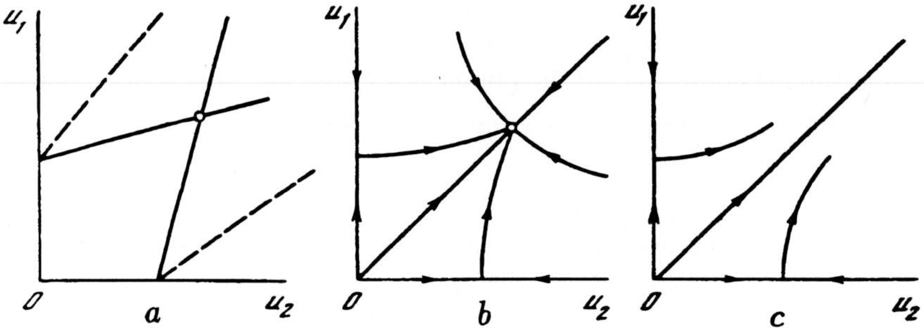

Fig. 4.2.1. Possible relative positions of the nullclines (a) for $p_1 p_2 < 1$ (solid lines) and for $p_1 p_2 > 1$ (dashed lines), and phase portraits (b and c) of system (4.2.2).

For $p_1 p_2 > 1$ the stable equilibrium does not exist, and for any nonzero initial condition both populations grow without bounds in spite of their intraspecies competition described by the logistic equations. The interpretation of this result lies in the mathematical description of the effect of protocooperation. The positive effect of protocooperation can compensate for the negative effect of intraspecies competition for all population sizes. This is clearly evidence for the limited range in which the bilinear description of protocooperation makes biological sense. We have already run into an analogous situation when we gave the bilinear description of predator–prey interaction. For system (4.2.1) the bilinear description of protocooperation means that an increase without bound of the population size of a partner leads to an unlimited increase of the reproduction rate of the other species, which is biologically absurd. The way out of this situation is obvious and analogous to the one we have used to describe the predator–prey system: protocooperation should be described by taking into account an effect similar to predator saturation. Thus, the system describing the dynamics of interacting populations takes the form

$$\begin{aligned}\dot{x}_1 &= a_1 x_1 \frac{K_1 - x_1}{K_1} + \frac{P_1 x_1 x_2}{1 + D_1 x_2}\,, \\ \dot{x}_2 &= a_2 x_2 \frac{K_2 - x_2}{K_2} + \frac{P_2 x_1 x_2}{1 + D_2 x_1}\,.\end{aligned} \tag{4.2.3}$$

The coefficients D_1 and D_2 are analogues of the saturation coefficients in (3.3.5). This implies that, when the population of, say, the second species grows without bound, the reproduction rate of the first species at a small population density increases, and asymptotically approaches $a_{\max} = a_1 + P_1/D_1$.

Setting $t = \tau/a_1$, $x_1 = K_1 u_1$ and $x_2 = K_2 u_2$ changes (4.2.3) into

$$\begin{aligned}\dot{u}_1 &= u_1(1 - u_1) + \frac{p_1 u_1 u_2}{1 + \delta_1 u_2}\,, \\ \dot{u}_2 &= \gamma u_2\left(1 - u_2 + \frac{p_2 u_1}{1 + \delta_2 u_1}\right),\end{aligned} \tag{4.2.4}$$

where $\delta_1 = D_1 K_2$, $\delta_2 = D_2 K_1$ and $\gamma = a_2/a_1$.

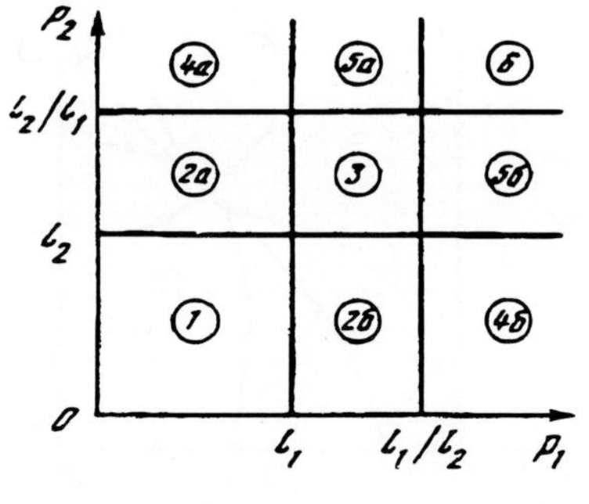

Fig. 4.2.2. The bifurcation diagram of system (4.2.6).

The relative positions of the nullclines are qualitatively identical for all parameters and correspond to the existence of a single globally attracting equilibrium in the phase portrait, which is like in Fig. 4.2.1*b*. This equilibrium corresponds to the coexistence of both populations. Coexisting populations have greater equilibrium sizes than isolated populations.

To conclude the discussion of protocooperation, we consider the case of two isolated populations with lower threshold sizes. Given the bilinear character of protocooperation, the dynamics of such a pair of populations is described by the system

$$\begin{aligned}\dot{x}_1 &= a_1 x_1 (x_1 - L_1)(K_1 - x_1) + P_1 x_1 x_2\,, \\ \dot{x}_2 &= a_2 x_2 (x_2 - L_2)(K_2 - x_2) + P_2 x_1 x_2\,.\end{aligned} \tag{4.2.5}$$

The standard coordinate change gives

$$\begin{aligned} \dot{u}_1 &= u_1(u_1 - l_1)(1 - u_1) + p_1 u_1 u_2 \,, \\ \dot{u}_2 &= \gamma u_2[(u_2 - l_2)(1 - u_2) + p_2 u_1] \,. \end{aligned} \tag{4.2.6}$$

The nullclines of the system are parabolas with minima located in the negative quadrants, and with branches pointing in the positive directions. In the first quadrant those parabolas may have two, three or four intersections. Figures 4.2.2 and 4.2.3 show the bifurcation diagram and a complete set of phase portraits of system (4.2.6), respectively.

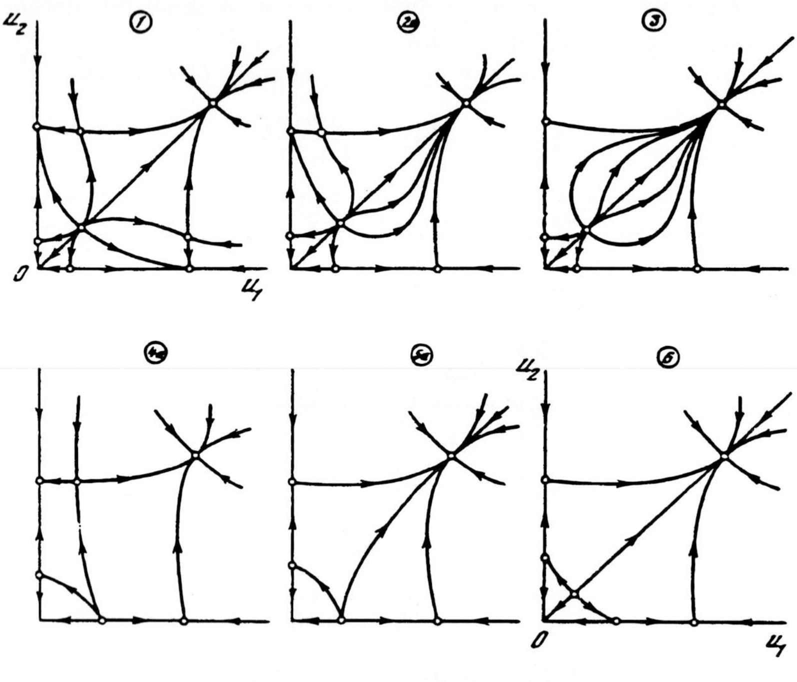

Fig. 4.2.3. Phase portraits of system (4.2.6). Exchanging u_1 and u_2 changes the phase portraits with index a to the one with index b.

The biological interpretation of the results is quite obvious. For all values of the parameters the state of coexisting populations and the trivial equilibrium are locally stable. It should be noted that the equilibrium densities of the coexisting populations are higher than their stable equilibrium densities in the absence of the partner. This is a manifestation of the positive effect of protocooperation.

We can say the following about the stability of an isolated population of one of the species when it is invaded by only a small number of individuals of the second species. Depending on the parameters, each of the isolated populations may be stable (region *1*), one of the isolated populations is stable (regions *2a*, *2b*, *4a* an *4b*), and none of the isolated populations is stable (regions *3*, *5* and *6*).

The main result of this is the following: with protocooperation the presence of a species-partner always leads to a reduction of the threshold size, and the threshold size may drop to zero with strong protocooperation.

4.2.2. *Mutualism*

In first approximation the dynamics of a pair of populations with mutualistic interaction can be described by the system

$$\begin{aligned} \dot{x}_1 &= -c_1 x_1 + P_1 x_1 x_2 \,, \\ \dot{x}_2 &= -c_2 x_2 + P_2 x_1 x_2 \,. \end{aligned} \tag{4.2.7}$$

Each of the populations is assumed to die out exponentially in the absence of a partner (in the same way as a predator does in the absence of prey). Mutualistic interactions are described by positive bilinear terms.

The nullclines of the system are straight lines parallel to the coordinate axes. The phase portrait of the system is qualitatively the same for all values of the parameters (Fig. 4.2.4*a*). The phase plane is divided into two regions separated by the stable manifold of the saddle *C*. For the initial conditions in the region adjoining the coordinate axes both populations die out. However, if the initial population densities are large enough then both populations grow without bounds. This fact is evidence of the inadequate character of a bilinear description of mutualism. (Something similar occured in the study of protocooperation.) Here we again make use of a factor similar to predator saturation, and system (4.2.7) takes the form

$$\begin{aligned} \dot{x}_1 &= -c_1 x_1 + \frac{P_1 x_1 x_2}{1 + D_1 x_2} \,, \\ \dot{x}_2 &= -c_2 x_2 + \frac{P_2 x_1 x_2}{1 + D_2 x_1} \,. \end{aligned} \tag{4.2.8}$$

It is easy to see that the picture does not change qualitatively, and that the nullclines remain straight lines parallel to the coordinate axes. Depending on the parameters, both populations either die out for any initial condition, or they die out only for some initial conditions and grow without bounds for other initial conditions. This is shown in Fig. 4.2.4*a*.

A more realistic description of the dynamics of two mutualistic populations may be obtained if we introduce the additional factor of intraspecies competition. This changes system (4.2.8) to

$$\begin{aligned} \dot{x}_1 &= -c_1 x_1 + \frac{P_1 x_1 x_2}{1 + D_1 x_2} - e_1 x_1^2 \,, \\ \dot{x}_2 &= -c_2 x_2 + \frac{P_2 x_1 x_2}{1 + D_2 x_1} - e_2 x_2^2 \,. \end{aligned} \tag{4.2.9}$$

If we exclude the trivial case of extinction of both populations for any initial conditions, the phase portrait of system (4.2.9) is always as shown in Fig. 4.2.4*b*.

The phase plane is divided into two regions separated by the stable manifold of the saddle C. For small initial population densities (points in phase space lying on the left and below the stable manifold) both populations are doomed to die out. For larger initial densities both populations eventually coexist in a stable equilibrium.

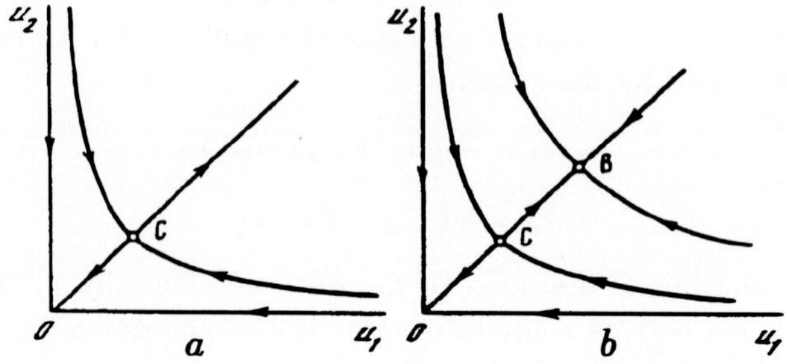

Fig. 4.2.4. Phase portraits of systems (4.2.7) (a) and (4.2.9) (b).

Both types of symbiotic interaction, protocooperation and mutualism, give rise to stable coexistence of the populations involved. For symbiosis between populations with lower threshold sizes the trivial state (both populations are extinct) is always stable. An isolated existence of one population (or both of them) may either be stable (against the invasion of the partner) or unstable, depending on how intense the symbiosis is. In other words, intense symbiosis overcomes the effect of a lower threshold size.

Appendix

A 4.1.1
Phase portraits of Eq. 4.1.3

$$\dot{x} = x(1 - x - \varepsilon_1 y)$$
$$\dot{y} = \gamma y(1 - \varepsilon_2 x - y)$$

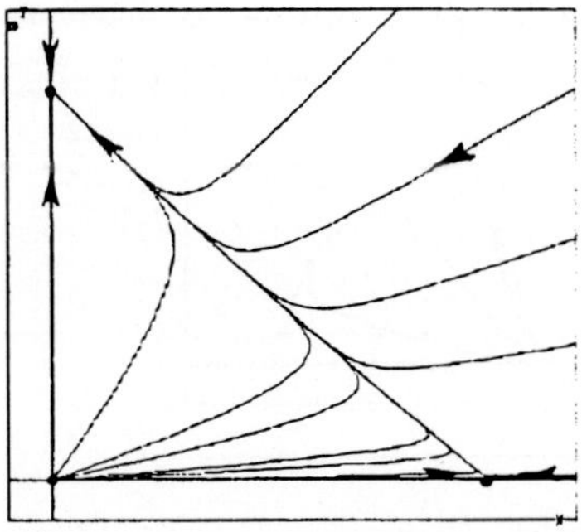

Region 2

$\gamma=1$
$\varepsilon_1=1.2$
$\varepsilon_2=0.9$

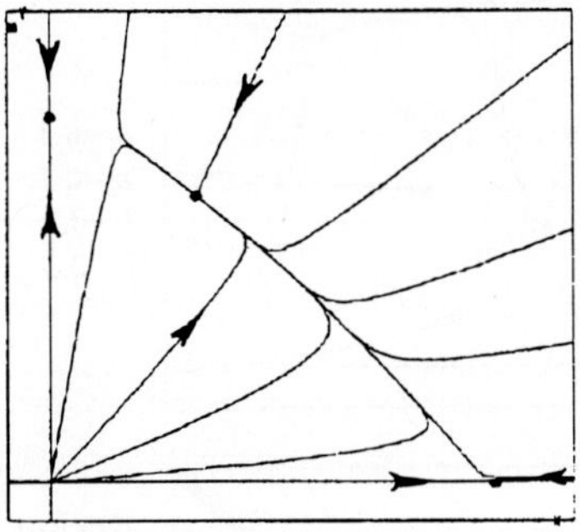

Region 3

$\gamma=1$
$\varepsilon_1=0.9$
$\varepsilon_2=0.8$

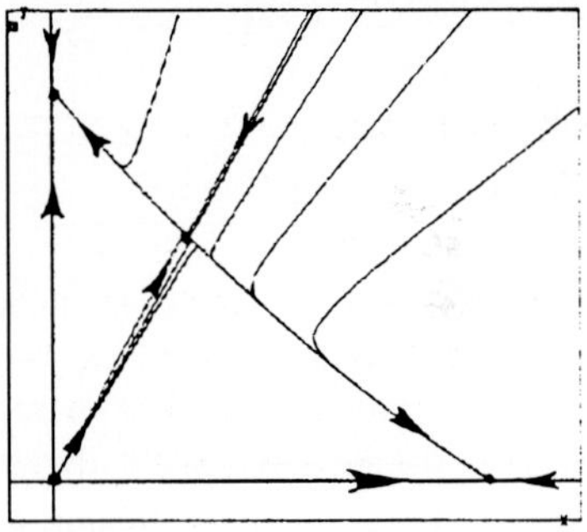

Region 4

$\gamma=1$
$\varepsilon_1=1.1$
$\varepsilon_2=1.2$

A 4.1.2
Phase portraits of Eq. 4.1.5

$$\dot{x} = x(x - l)(1 - x) - \varepsilon_1 xy$$
$$\dot{y} = \gamma y(1 - \varepsilon_2 x - y)$$

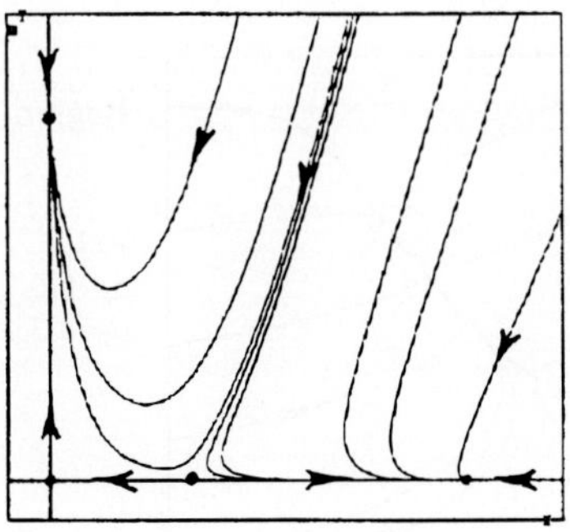

Region 1

$\gamma=1$
$\varepsilon_1=1$
$\varepsilon_2=3.5$
$l=0.3$

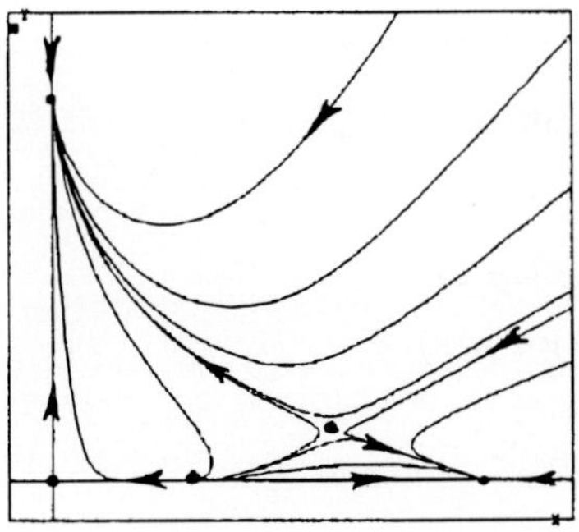

Region 2

$\gamma=1$
$\varepsilon_1=1$
$\varepsilon_2=1.3$
$l=0.3$

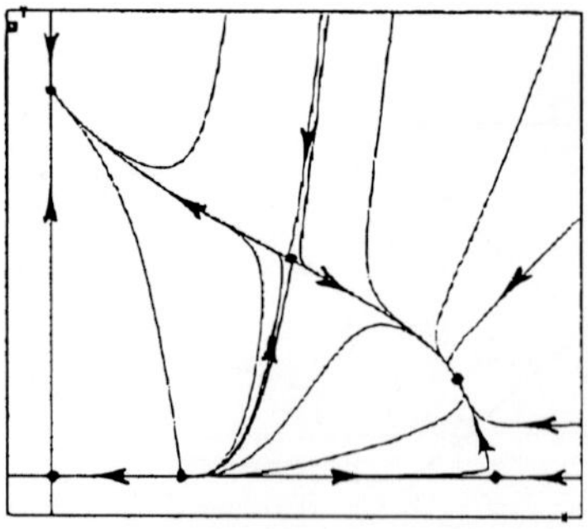

Region 3

$\gamma=1$
$\varepsilon_1=0.2$
$\varepsilon_2=0.8$
$l=0.3$

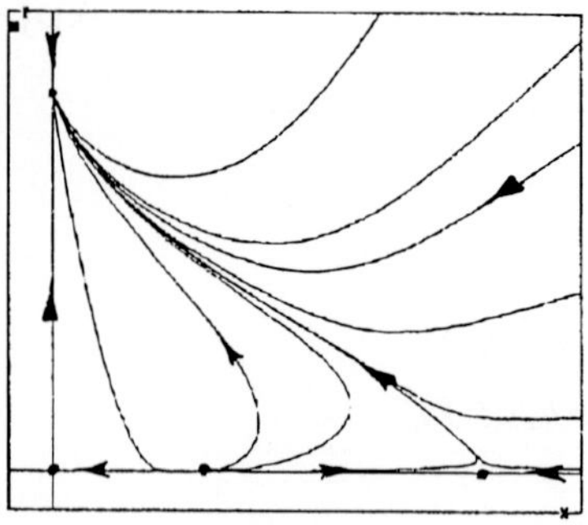

Region 4

$\gamma=1$
$\varepsilon_1=0.7$
$\varepsilon_2=0.8$
$l=0.3$

A 4.1.3
Phase portraits
of Eq. 4.1.7

$$\dot{x} = x(x - l_1)(1 - x) - \varepsilon_1 xy$$
$$\dot{y} = \gamma y\left[(y - l_2)(1 - y) - \varepsilon_2 x\right]$$

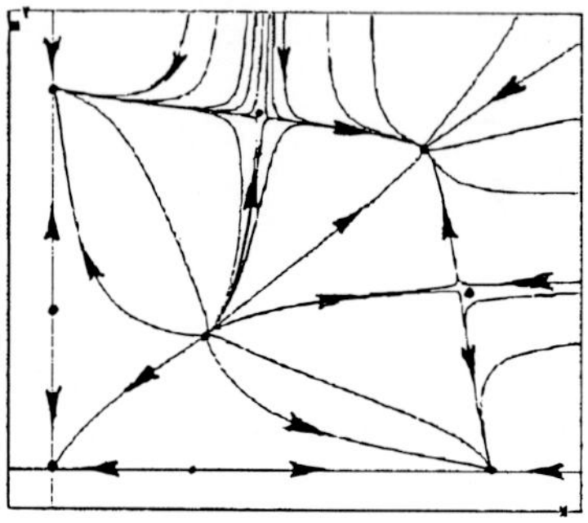

Region 1

$\gamma=1$
$\varepsilon_1=0.1$
$\varepsilon_2=0.1$
$l_1=0.3$
$l_2=0.3$

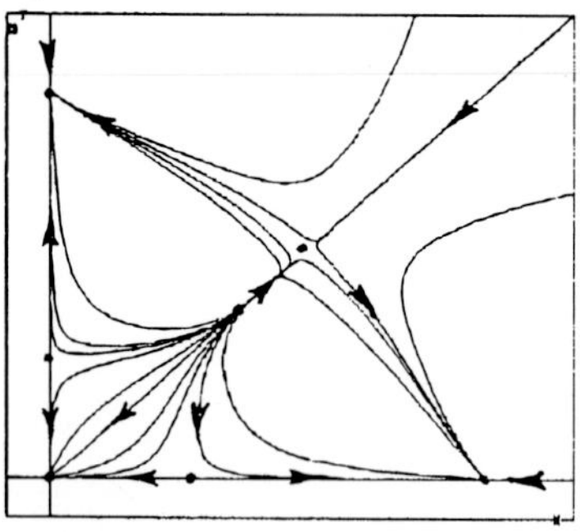

Region 2

$\gamma=1$
$\varepsilon_1=0.2$
$\varepsilon_2=0.2$
$l_1=0.3$
$l_2=0.3$

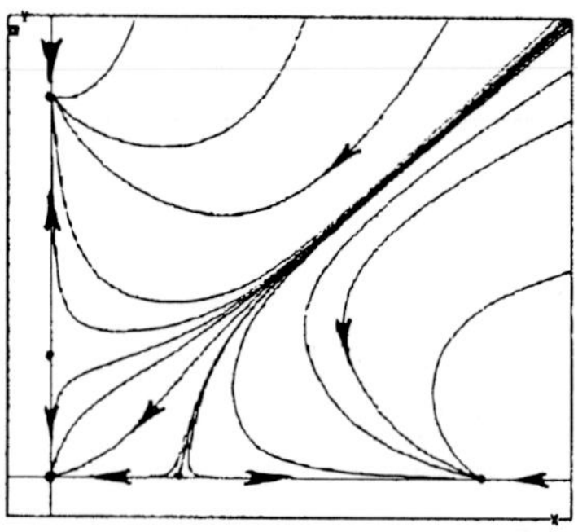

Region 3

$\gamma=1$
$\varepsilon_1=0.3$
$\varepsilon_2=0.3$
$l_1=0.3$
$l_2=0.3$

A 4.2.1
Phase portrait
of Eq. 4.2.2

$$\dot{x} = x(1 - x + p_1 y)$$
$$\dot{y} = \gamma y(1 - y + p_2 x)$$

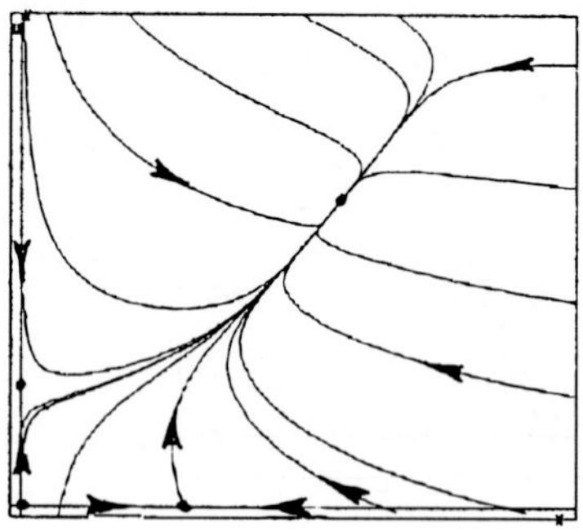

Region 1

$\gamma=1$
$p_1=0.6$
$p_2=0.7$

Phase portraits of Eq. 4.2.6

$$\dot{x} = x(x - l_1)(1 - x) + p_1 xy$$
$$\dot{y} = \gamma y\left[(y - l_2)(1 - y) + p_2 x\right]$$

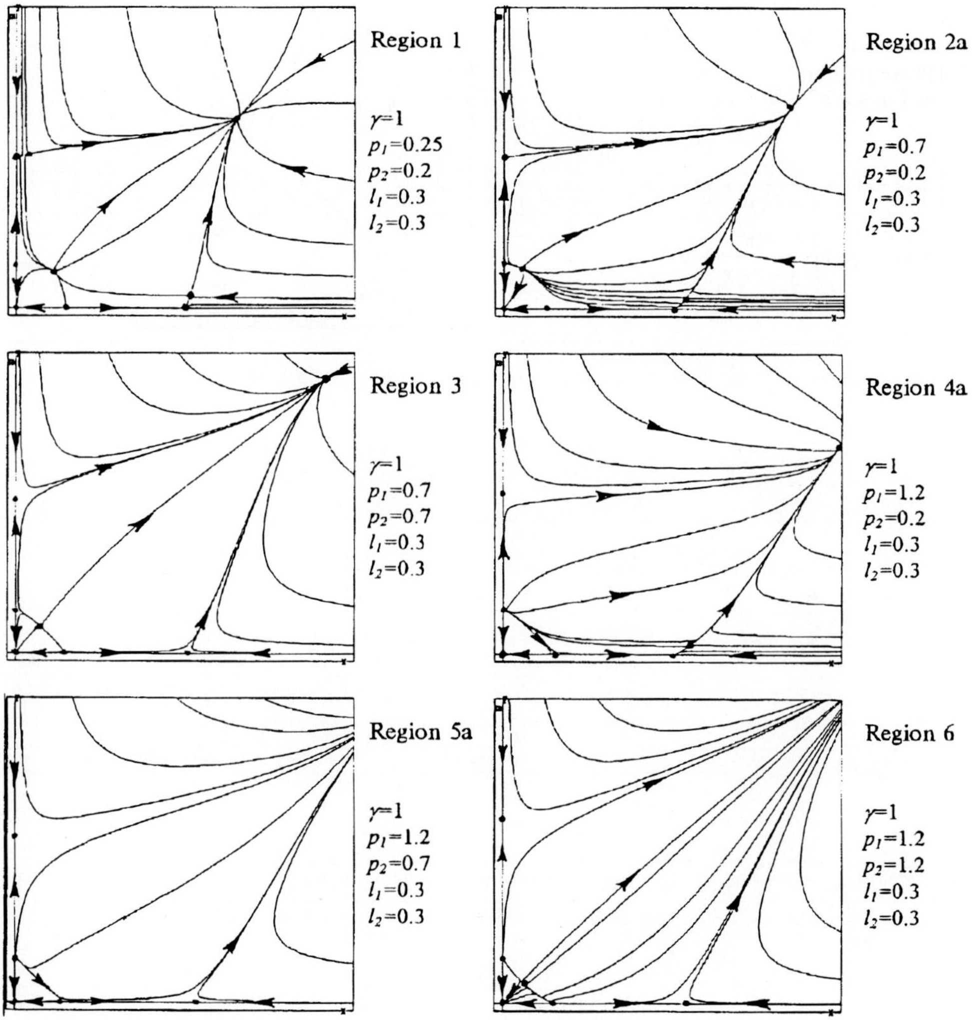

A 4.2.2
Phase portrait
of Eq. 4.2.7

$$\dot{x} = -x(1-y)$$
$$\dot{y} = -c_2 y(1-x)$$

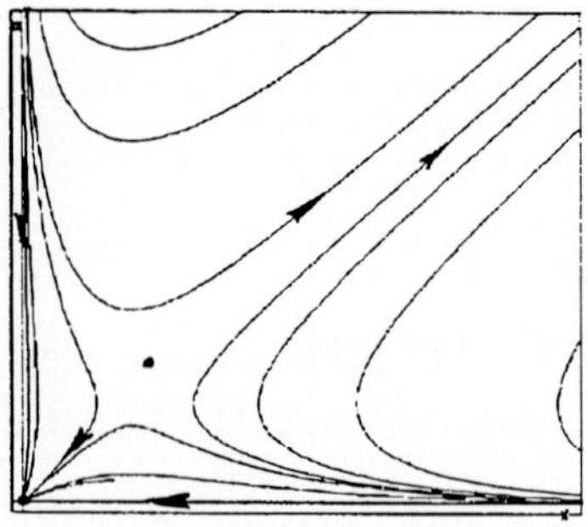

$c_2=1$

Special case:
$c_1=p_1=1$
$p_2=c_2$

Phase portrait
of Eq. 4.2.9

$$\dot{x} = -x\left(1-\frac{P_1 y}{1+D_1 y}+x\right)$$
$$\dot{y} = -c_2 y\left(1-\frac{P_2 x}{1+D_2 x}+y\right)$$

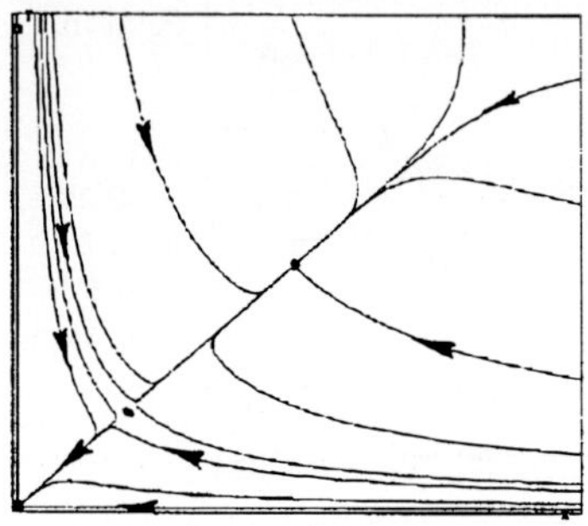

$c_2=1$
$P_1=1.8$
$P_2=1.8$
$D_1=0.1$
$D_2=0.1$

Special case:
$c_1=e_1=1$
$c_2=e_2$

Chapter 5

Local Systems of Three Populations

In Chapter 3 and Chapter 4 of this book we showed that for *a pair* of interacting populations it is possible to give a reasonably complete classification of factors that should be taken into account in the analysis of the dynamics. Wherever possible, we gave a complete analysis of the dynamical consequences to which the separate factors and their various combinations can lead. At present, it is unrealistic, or at least premature, for a number of reasons to make an analogous study of model ecosystems consisting of *three or more* populations.

First, it is difficult to review various possible models consisting of three populations, even if we allow only for the factors analysed in the previous chapters.

Second, the qualitative theory of differential equations, which is mainly used here and which, in a sense, may be regarded as complete for two-dimensional systems is far from completeness in the case of three and higher-dimensional systems. In particular, it leaves open the question about possible strucures of attractors and their classification in phase spaces of dimension higher than two.

The third point is very much related to the second. The systems of differential equations describing the dynamics of three or more interacting populations contain large numbers of parameters even in their scaled forms. Earlier it has been said that the analysis of neighborhoods in parameter space of bifurcations of higher, and in the ideal case, highest possible, codimensions (for the given system), is the most convenient technique for a complete qualitative study (Molchanov, 1975*b*). Thus, in order to study three-dimensional systems depending on a large number of parameters, it is necessary to know the normal forms of bifurcations of higher codimensions in three-dimensional phase space. The normal forms for many bifurcations of codimension more than one are not known.

At present, it is hardly possible to give a complete classification of the dynamic regimes and their transformations in ecological models including more than two populations. This is why we shall dwell in this chapter on the study of trophic interrelations in systems of three interacting populations with and without inter- and intraspecies competition. The emphasis will be on dynamic regimes that are realized in local systems of three interacting populations, but are absent in systems of two interacting populations. Different aspects of dynamical regimes of multi-dimensional modifications of Volterra's model were analysed by Nakajima and De Angelis (1989), Busenberg *et al.* (1990), and other.

5.1. Classification of Trophic Structures

It has been shown above that populations can mainly interact in three ways: competition, symbiosis, and predator–prey (or trophic) relations (Odum, 1971). In this section we concentrate on trophic relations and give a classification of the trophic structures that are possible in a system of three interacting populations.

We represent the populations by vertices of a graph, and the trophic relations between them by arrows indicating the directions of the flow of a substance. Obviously, there are only two types of trophic graphs proper of such a system (Fig. 5.1.1). We call them a cycle (left) and a cell (right) of the network.

Fig. 5.1.1. The two types of trophic relations in a system of three populations.

These graphs only represent the interpopulation relations in the system. In order to have a view of the functioning of the system, we need to know in addition how each of the three populations forming the ecosystem behaves if it is left to itself. In other words, we must know which populations are autotrophic, and which are heterotrophic, that is, which populations, if left to themselves, grow and which die out. It is natural to represent an autotrophic relation by an arrow coming into a vertex of the graph and heterotrophic relations by arrows coming out of a vertex.

If we list all possible combinations of autotrophic and heterotrophic interactions we easily see that a cycle gives four types of trophic structures, whereas a cell gives eight types. In total, a system of three interacting populations can formally be of twelve trophic types.

Now we model each trophic structure of Fig. 5.1.2 with a Volterra system of third-order differential equations by assuming that the incoming and outgoing arrows represent linear terms, and that the arrows connecting pairs of populations correspond to bilinear terms. For example, graph a corresponds to the system

$$\begin{aligned} \dot{x} &= a_1 x - b_{12} xy - b_{13} xz \,, \\ \dot{y} &= a_2 y + d_{12} xy + d_{23} yz \,, \\ \dot{z} &= a_3 z - b_{23} yz + d_{13} xz \,. \end{aligned} \tag{5.1.1}$$

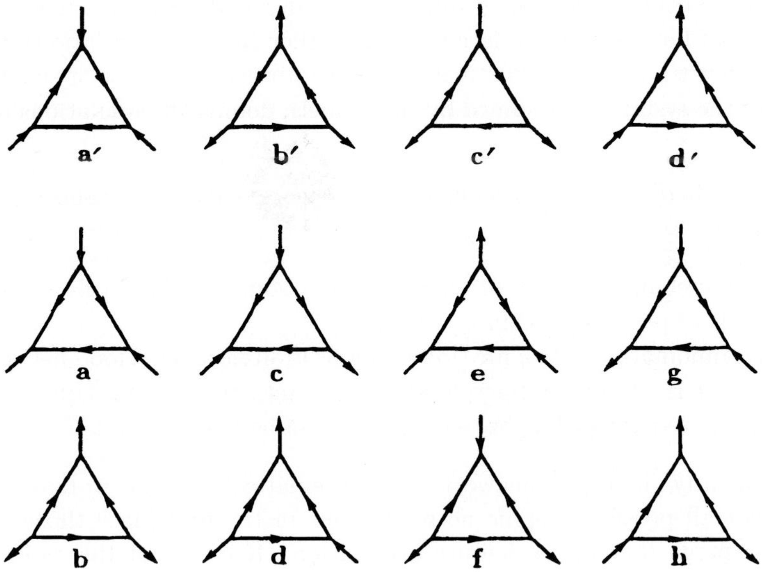

Fig. 5.1.2. Trophic structures of cycle-type (a'–d') and of cell-type (a–h).

Before turning to the differential equations describing the trophic structures in Fig. 5.1.2, we give their biological interpretation. Our aim is to exclude most trophic structures from further consideration.

Remark 1. If we change the direction of arrows in Fig. 5.1.2 the graphs a' and c' transform into graphs b' and d', respectively, and the graphs of the middle row transform into their counterparts in the lower row (and vice versa). This means that changing the direction of time converts the corresponding systems of differential equations into each other. Reversing the direction of time means passing from one scheme of trophic relations to another with opposite directions of all substance–energy flows. In that case, autotrophs become heterotrophs, that is, the prey becomes the predator and vice versa. Recall that Volterra's predator–prey model (3.1.1) is invariant (of course up to scalings of the parameters) under time reversal.

Remark 2. For systems corresponding to the graphs a and a' in Fig. 5.1.2 there exists a general theorem (Volterra, 1931): if all parameters of the linear terms are positive, then the trajectories of the system go to infinity for any initial condition.

An analogous statement is available for the systems presented by the graphs b and b': if all parameters of the linear terms are negative, then the origin of the phase space is a globally attracting equilibrium.

The interpretation is obvious. The graphs a' and a describe ecological models of three autotrophs connected by trophic interrelations. Here every component of the system left to itself grows without bound. Naturallly, so does the system as a whole. On the other hand, the graphs b' and b describe ecological models of three populations of heterotrophs, related to each other by trophic relationships. Each population left to itself dies out. Naturally, the absence of inflow of substance and energy into the system, as assumed by the models, dooms all populations to become extinct.

Remark 3. There are several idealizations in theoretical ecology concerning trophic relationships:

(a) trophic chain: transfer of energy from a source of food (plants) occurs by means of predation (Odum, 1971).
(b) the combination of trophic (food) chains is often called a food network (ibid).
(c) organisms obtaining their food from plants through an equal number of stages are regarded as belonging to the same trophic level.

Consequently, in the framework of mathematical ecology it makes sense to consider not all possible trophic networks, but in the first place those where the substance–energy flow has a definite orientation. If we adopt this point of view, then two types of possible trophic networks should be excluded from further consideration. First, we exclude networks in whose graphs some of the vertices have only incoming or only outgoing substance–energy flows. The reason for this seems quite obvious. Second, networks containing cycles should be excluded, too. The reason for this is twofold. First, the flow of substance in the ecosystems is not unidirectional, but forms a number of circulations. It goes without saying that the dynamics of such circulations is for a large part specified by the presence of abiotic substances. The consideration of such systems is beyond the scope of this book; they were studied in detail by V.V. Alexeev and his colleagues (Alexeev, 1976; Alexeev *et al.*, 1978; Polyakova, Sazykina, 1976). Second, it seems that the examples of oppositely directed or closed cyclic flows of substance in trophic networks can be found in nature (carnivorous plants of the sundew type etc.). But such phenomena are the exceptions rather than the rule and, therefore, should be considered as "second-order effects".

Let us now consider which of the trophic structures shown in Fig. 5.1.2 meet the requirements formulated above and deserve further consideration. Obviously, the structures a'–d' are excluded because they contain cycles. Of the structure of the middle and the lower row we exclude the graphs a, c, and e (containing vertices with three incoming substance–energy flows) and graphs b, d and f (containing vertices with three outgoing flows).

Thus, of all possible trophic graphs of a three-population community, only the graphs g and h, which change into each other by reversing time, are oriented trophic networks.

Terminological remark. When describing trophic relations in a system of three populations it is reasonable to use the term *producer*, and to keep the term *prey*, for a species being the food for two other populations of the community. In the same way, we keep the term *predator* for a species consuming both of its partners in the community. The term *consumer* can be employed for a species that is itself a predator of prey, but also prey for a predator.

What is the possible ecological interpretation of graphs g and h shown in Fig. 5.1.2? This interpretation becomes obvious if we modify the figures so that the auto- and heterotrophs are at different vertical levels (Fig. 5.1.3*a*). Graph g shows two autotrophs and one heterotroph, with one of the two autotrophs being a facultative predator of the other. Graph g shows one autotroph and two heterotrophs, with one of the heterotrophs being a facultative predator of the other.

Note that the terminology and the graphic methods of representation used in this field are not quite agreed on: it is accepted to consider predators as the upper trophic level, whereas it is common to depict the flow of substance and energy as going from top to bottom. If we locate the components of the analysed model at three trophic levels (with predator on the top) then the graphs shown in Fig. 5.1.3*a* take the form in Fig. 5.1.3*b*.

Above, we have excluded the exotic situations when the system has an autotrophic predator. Thus, the trophic graph shown in Fig. 5.1.3*a* is also out of the scope of our analysis. On the other hand, the trophic graph h of Fig. 5.1.3*b*, which corresponds to the facultative "herbivourousness" of the predator, describes a very common ecological situation and will be analysed in detail.

Until now, we have only considered complete trophic graphs with all possible relations (of one or the other sign) between populations. However, trophic graphs where some relations are absent are of interest, too. Obviously, only some of the theoretically possible trophic relations occur in real ecological systems. In a system of three interacting populations our interest lies in three types of degenerate (incomplete) trophic structures. In the absence of one of the trophic relations they are particular cases of complete trophic structures: one-predator–two-preys (populations), two-predators–one-prey, and producer–consumer–predator without facultative herbivourousnes of the predator (Fig. 5.1.4, graphs *a*–*c*).

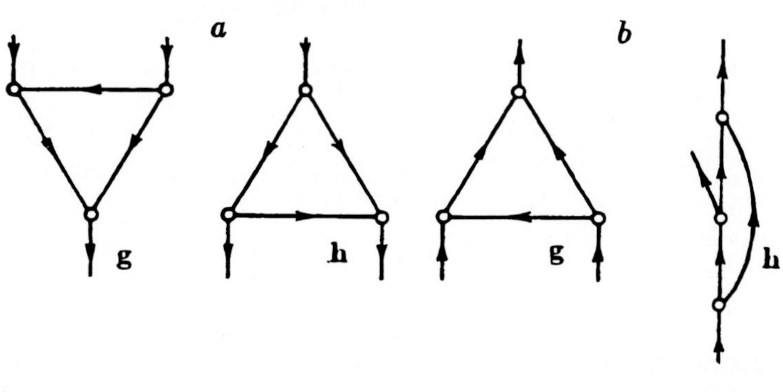

Fig. 5.1.3. Trophic graphs (a): two-autotrophs–one-heterotroph (g), one-autotroph–two-heterotrophs (h), and their alternative representation (b) in the case of facultative predation.

The first two cases are exchanged by reversing the direction of time. We shall start with them in our study of the differential equations describing systems of three interacting populations.

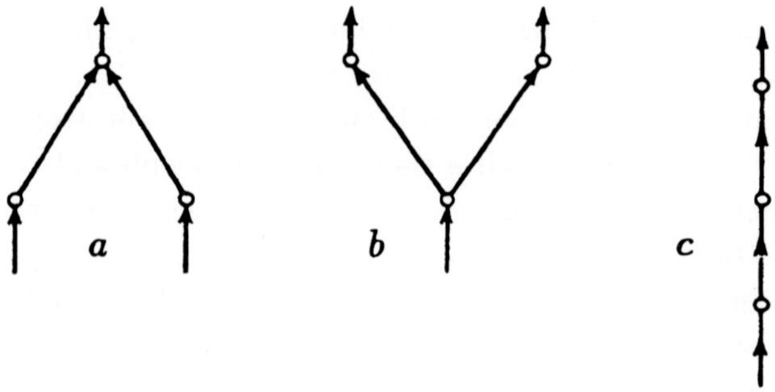

Fig. 5.1.4. The three most interesting cases of degenerate trophic structures in a three-population system.

5.2. Competition-Free Communities

Let us consider the differential equations that describe systems of three populations involved in trophic relations corresponding to the graphs which we have analysed in the previous section. It is assumed that the trophic relations between populations are in quantitative agreement with Volterra's scheme, so that they can be described by bilinear terms, while all other, in particular competitive interactions, are absent.

5.2.1. *One-Predator–Two-Preys and Two-Predators–One-Prey*

The trophic graph a (Fig. 5.1.4) can be described by the following system of differential equations

$$\begin{aligned} \dot{x}_1 &= a_1 x_1 - b_1 x_1 y\,, \\ \dot{x}_2 &= a_2 x_2 - b_2 x_2 y\,, \\ \dot{y} &= -cy + d_1 x_1 y + d_2 x_2 y\,. \end{aligned} \tag{5.2.1}$$

Setting $t = \tau/a_1$, $x_1 = (c_1/d_1)u_1$, $x_2 = (c_2/d_2)u_2$, and $y = (a_1/b_1)v$ gives

$$\begin{aligned} \dot{u}_1 &= u_1(1 - v)\,, \\ \dot{u}_2 &= \gamma_1 u_2(n - v)\,, \\ \dot{v} &= -\gamma_2 v(1 - u_1 - u_2)\,, \end{aligned} \tag{5.2.2}$$

where $\gamma_1 = b_2/b_1$, $\gamma_2 = c/a_1$ and $n = a_2 b_1/a_1 b_2$.

The coordinate planes of the system are invariant. For $n \neq 1$ the system has no nontrivial (that is, not lying in a coordinate plane) equilibria. The equilibria, other than the origin, are:

$$\mathbf{A}_1(u_1 = 1,\ u_2 = 0,\ v = 1)\,,$$

$$\mathbf{A}_2(u_1 = 0,\ u_2 = 1,\ v = n)\,.$$

In the (u_1, v)- and (u_2, v)-planes the points A_1 and A_2 are centers while the respective trajectories form one-parameter families of closed orbits.

Without loss of generality we may assume that $n > 1$. Then the (one-dimensional) unstable manifold of the point A_1 and also the (one-dimensional) stable manifold of the point A_2 both enter the first phase octant (Fig. 5.2.1). (For $n < 1$ the situation is reverse.) For $n = 1$, the phase portrait exhibits the straight line of non-isolated equilibria $\{u_1 + u_2 = 1,\ v = 1\}$.

In order to understand how the trajectories of the system are arranged globally, we consider the case when $\gamma_1 = 1$ and change to cylindrical coordinates: $u_1 = \rho\cos\varphi$, $u_2 = \rho\sin\varphi$.

Now system (5.2.2) can be written for $\gamma_1 = 1$ as

$$\begin{aligned}
\dot\rho &= \rho\left(1 - v + (n+1)\sin^2\varphi\right)\,,\\
\dot\varphi &= 1/2(n-1)\sin 2\varphi\,,\\
\dot v &= -\gamma v\,(1 - \rho(\sin\varphi + \cos\varphi))\,.
\end{aligned} \tag{5.2.3}$$

It follows from (5.2.3) that the invariant plane $\{\varphi = 0\}$ (formerly (u_1, v)) is repelling as a whole, while the plane $\varphi = \pi/2$ (formerly (u_2, v)) is attracting.

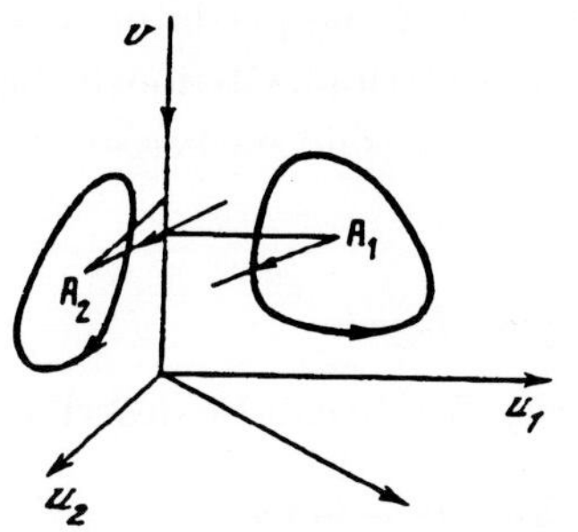

Fig. 5.2.1. Phase portrait of system (5.2.2).

Now we can imagine the structure of the phase space. The unstable manifold of A_1 enters the first octant and is attracted to a closed orbit lying in the (u_2, v)-plane. Correspondingly, the stable manifold of A_2 comes from the first octant and from a closed orbit in the (u_1, v)-plane. In particular, trajectories winding off into the octant from an arbitrary closed orbit in the (u_1, v)-plane do not wind onto any definite closed orbit in the (u_2, v)-plane. Instead they go to a set of trajectories bounded by some annulus in this plane. Similarly, the trajectories attracted to an arbitrary closed orbit in the (u_2, v)-plane wind off from a set of closed orbits in the (u_1, v)-plane bounded by some annulus in this plane. In other words, the trajectories behave as if they are mixed in the interior of the octant. The situation is not structurally stable: if we allow for some other factor of population dynamics, which would be presented by new terms in the right-hand sides of (5.2.1), the system may change its behavior qualitatively. Nevertheless, the fact that the trajectories can be mixed, which has no analogy in two-dimensional dynamical systems, as we shall see below, is a property of system (5.2.1) as well as of some of its structurally stable modifications as well.

The biological interpretation of the results of the study of system (5.2.1) is obvious. When the interactions of each of the two prey populations with the predator population are bilinear and there are no other factors affecting the dynamics, one of the prey populations always outcompetes the other. In this contest the population wins that can ensure a higher stationary density of the predator population. This may be due to, say, a higher biotic potential of one of the prey populations, when their other characteristics are equal. The amplitude of oscillations in the system as one population is outcompeting the other may increase as well as decrease.

The system of two predators and one prey, described by the differential equations

$$\begin{aligned} \dot{x} &= ax - b_1 x y_1 - b_2 x y_2 \,, \\ \dot{y}_1 &= -c_1 y_1 + d_1 x y_1 \,, \\ \dot{y}_2 &= -c_2 y_2 + d_2 x y_2 \,, \end{aligned} \tag{5.2.4}$$

behaves in quite an analogous manner. One of the predator populations always drives out the other. In this contest the predator wins that ensures the minimum stationary density of the prey population. Obviously, here the trajectories are mixed because system (5.2.4), as has been said earlier, changes into (5.2.1) by reversing the direction of time.

5.2.2. *System of Three Trophic Levels*

The trophic graph shown in Fig. 5.1.4*c* can be described by the system of equations

$$\begin{aligned} \dot{x} &= ax - b_1 xy \,, \\ \dot{y} &= -c_1 y + d_1 xy - b_2 yz \,, \\ \dot{z} &= -c_2 z + d_2 yz \,. \end{aligned} \tag{5.2.5}$$

Setting $t = \tau/a$, $x = (c_1/d_1)u$, $y = (a/b_1)v$, and $z(c_1/b_2)w$ changes this to

$$\begin{aligned} \dot{u} &= u(1 - v) \,, \\ \dot{v} &= -\gamma_1 v(1 - u + w) \,, \\ \dot{w} &= -\gamma_2 w(\alpha - v) \,, \end{aligned} \tag{5.2.6}$$

where $\gamma_1 = c_1/a$, $\gamma_2 = d_2/b_1$, and $\alpha = c_2 b_1/ad_2$.

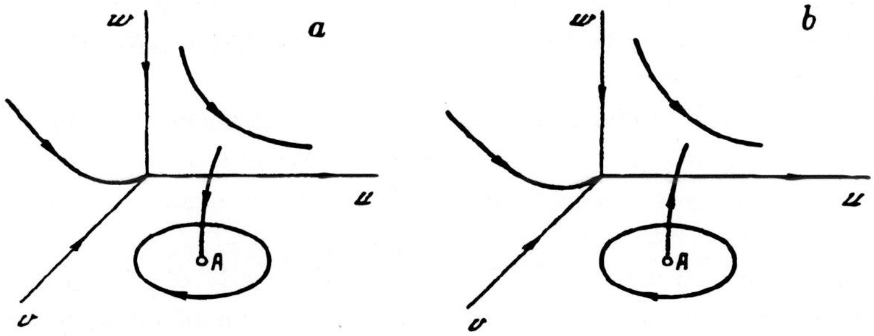

Fig. 5.2.2. Phase portraits of system (5.2.6) for $\alpha > 1$ (a), and for $\alpha < 1$ (b).

Equilibria of the system are the origin and the point $A(u = v = 1, w = 0)$. The origin is obviously a three-dimensional saddle point, where the u-axis is the unstable direction and the other coordinate axes are the stable directions. The point A in the (u, v)-plane is a center surrounded by a one-parameter family of closed orbits. Its stability in the w-direction is given by the sign of the quantity $\dot{w}/w|_{\mathbf{A}} = -\gamma_2(\alpha - 1)$. The phase portraits of system (5.2.4) for $\alpha > 1$ and for $\alpha < 1$ are shown in Fig. 5.2.2. For $\alpha = 1$ (that is, at the bifurcation value) the system has the straight line of non-isolated fixed points $\{u = 1 + w,\ v = 1\}$.

Biologically this can be interpreted as follows. The consumer population density v in a system of three trophic levels is regulated by two factors: by the size of the producer population and by predator pressure. If the equilibrium density of the consumer population in the absence of the predator is less than its density ensuring stationary existence of the predator ($\alpha > 1$) then the predator is doomed to die out, and the consumer and producer populations regulate their mutual sizes.

In the other case, the predator so much overwhelms the consumer population that the producer population escapes the control on part of the consumer and starts to grow without bound. The predator population then also grows without bound.

This incorrectness of model (5.2.5), manifesting itself in the possibility of an unlimited growth of the predator population, is caused by two circumstances. The first is the assumption that the producer population should grow exponentially in the absence of a consumer. The second assumption is that the interactions between the producer and the consumer, as well as between the consumer and the predator are considered to be bilinear. As a result of this, it turns out that a *limited* population of the consumer can absorb the producer at an *unlimited* rate, converting it into its own biomass and ensuring, thereby, the unlimited growth of the predator population. It is clear that both these assumptions are only valid within a certain range of population densities.

5.2.3. *Cell of a Trophic Net*

In the previous sections of this chapter we have considered the systems of differential equations corresponding to incomplete trophic graphs with one of the trophic relations being absent. Now we shall consider the complete trophic graph d of the cell of a trophic net shown in Fig. 5.1.3*b*. It can be described by the system of differential equations

$$\begin{aligned} \dot{x} &= ax - b_1xy - b_3xz\,, \\ \dot{y} &= -c_1y + d_1xy - b_2yz\,, \\ \dot{z} &= -c_2z + d_2yz + d_3xz\,. \end{aligned} \tag{5.2.7}$$

Setting $t = \tau/a$, $x = (c_1/d_1)u$, $y = (a/b_1)v$, and $z = (c_1/b_2)w$ changes this to

$$\begin{aligned} \dot{u} &= u(1 - v - \beta w)\,, \\ \dot{v} &= -\gamma_1 v(1 - u + w)\,, \\ \dot{w} &= -\gamma_2 w(\alpha - v - \delta u)\,, \end{aligned} \tag{5.2.8}$$

where $\gamma_1 = c_1/a$, $\gamma_2 = d_2/b_1$, $\alpha = b_1c_2/ad_2$, $\beta = c_1b_3/ab_2$, and $\delta = b_1c_1d_3/ad_1d_2$.

The equilibria of the system are

$$\mathbf{O}(u = v = w = 0)\,; \quad \mathbf{A}_1(u = v = 1,\ w = 0)\,;$$

$$\mathbf{A}_2(u = \alpha/\delta,\ v = 0,\ w = 1/\beta)\,;$$

$$\mathbf{A}_3\left(u = \frac{1 + \beta - \alpha}{\beta - \delta},\ v = \frac{\alpha\beta - \delta(\beta + 1)}{\beta - \delta},\ w = \frac{1 - \alpha + \delta}{\beta - \delta}\right).$$

Note that (5.2.8) represent the first of the "non-Volterra" systems considered in this chapter. Indeed, although the terminology used in this field is not quite agreed on, Volterra systems are usually thought to satisfy the following requirements: trophic relations should be described by bilinear terms, and the matrix of parameters describing the trophic relations should have a sign-symmetric form. The principle of *equivalents* should be observed according to which $d_{rs} = -d_{sr}$ for systems of the type $\dot{x}_r = x_r(a_r + (1/b_r)\sum_s d_{rs}x_s)$.

For system (5.2.8) the above condition means that $\beta = \delta$. However, we are interested in the general case when $\beta \neq \delta$.

Let us examine the bifurcations of the equilibria of the system. As always, we limit ourselves to the positive octant of the phase space. There are two bifurcations of this type: saddle-node bifurcations of the nontrivial equilibrium A_3 (lying within the octant) with the equilibria A_1 and A_2, respectively. The equations defining the corresponding bifurcation surfaces in parameter space can be written as

$$\begin{aligned} a_{13} &: \delta = \alpha - 1\,; \\ a_{23} &: \delta = \alpha\beta/(\beta + 1)\,. \end{aligned}$$

A complete idea of the structure of the bifurcation diagram in (α, β, δ)-space, with respect to bifurcations occuring in the first octant of the phase space, can be given by a family of cross sections, namely (α, δ)-planes for all values of β (Fig. 5.2.3*a*). The dotted straight lines show the bifurcations that occur outside the first octant. The phase portraits of the system for the corresponding parameter regions are shown in Fig. 5.2.3*b*.

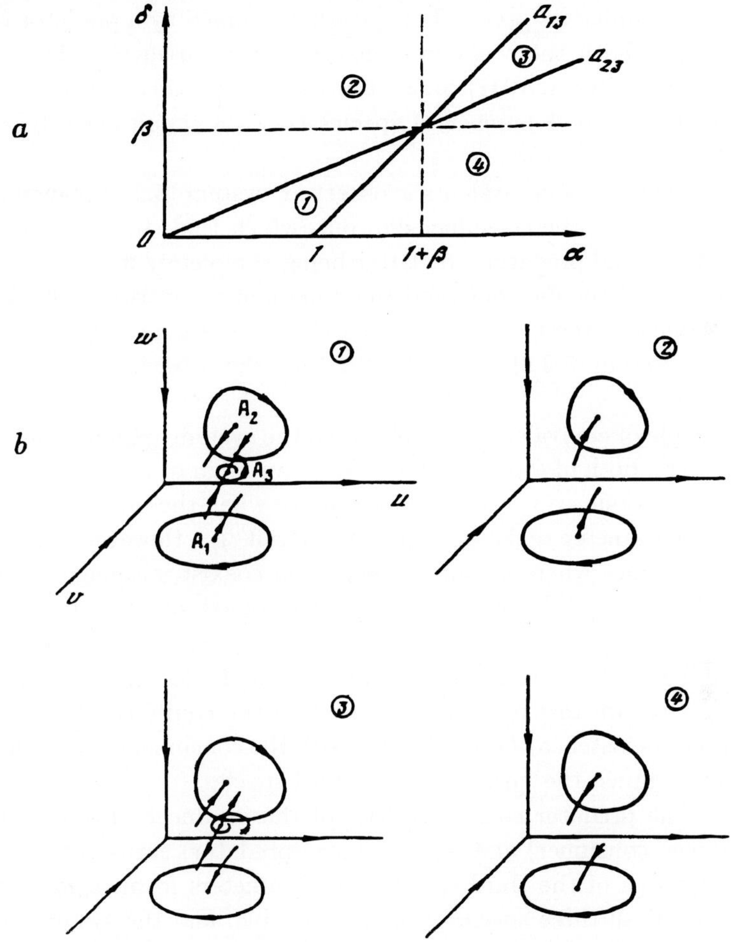

Fig. 5.2.3. Bifurcation diagram (*a*) and phase portraits (*b*) of system (5.2.8).

There are no nontrivial equilibria for parameters from regions *2* and *4*. For parameters from regions *1* and *3* a nontrivial equilibrium exists, though it is unstable. Its type of instability depends on where the parameters lie: in region *1* we have the condition $\lambda_1 < 0$, Re $\lambda_2 > 0$, Re $\lambda_3 > 0$, and in region *3* we have the contrary, namely $\lambda_1 > 0$, Re $\lambda_2 < 0$, Re $\lambda_3 < 0$. Note that the real eigenvalue λ_1 and the real part of the complex pair Re λ_2 and Re λ_3 are both zero at the point in the

(α, β)-plane where the bifurcation curves a_{13} and a_{23} intersect. In other words, this bifurcation has conditional codimension one on account of the species type of the system.

What is the biological interpretation of these results? To make this clear, we first examine the special case of system (5.2.8) when the principle of equivalents $\beta = \delta$ is satisfied. Here, the biological interpretation of this principle is based on the assumption that there should exist a constant biomass conversion-coefficient for every species consuming its prey or being itself consumed by a predator (Svirezhev, 1976). Note that both these assumptions are rather artificial. The well-known general theorem (Volterra, 1931) for systems satisfying this principle says that it is impossible that an odd number of species coexists simultaneously, even in an unstable regime.

When the principle of equivalents is observed, system (5.2.8) may evolve in two ways. First, for $\alpha < 2$, the consumer dies out, which leads to a community of two species – producer and predator (the latter being completely herbivorous). Second, for $\alpha > 2$, the predator dies out, and the community consisting of producer and consumer populations remains. Thus, when observing the principle of equivalents, the behavior of system (5.2.8) does not differ from the behavior of a two-predators–one-prey system.

If the principle of equivalent is not observed the system exhibits, for some values of the parameters, both of the above regimes for any sign of $\beta - \delta$. For large enough α the predator dies out, as it was described above when the principle of equivalents was observed (parameter region and phase portrait *4*). However, for $\beta \neq \delta$ there are parameter regions where all three species can coexist, although in an unstable regime. The behavior of the system here is determined, on the whole, by the sign of $\beta - \delta$.

For $\beta < \delta$ the predator consumes a comparatively large amount of the producer. As a result, for parameters from region *3* each of the equilibria A_1 and A_2, corresponding to the coexistence of the producer with the consumer or with the predator, becomes stable against the invasion of the third species.

For $\beta > \delta$ the predator consumes little of the producer. Each of the equilibria A_1 (producer–consumer) and A_2 (producer–predator) turns out to be unstable against the invasion of the third species for parameters from region *1*. However, the coexistence of all three species remains unstable, and the system gives rise to oscillations with an ever increasing amplitude.

One of the reasons of this incorrectness of the model is obviously the assumption that the producer population grows exponentially and without bound in the absence of a consumer, which is the assumption that the resources of the producer population are unlimited. In other words, there is no intraspecies competition in the producer population.

The analysis of system (5.2.8) has revealed that the rejection of the biologically unsound principle of equivalents may give rise to a qualitatively new behavior of a simple trophic cell. In particular, this can be a coexistence, although an unstable one, of the three species of the cell. This result makes the assumption of the principle of equivalents questionable when one studies more complicated trophic networks.

5.3. Competing Producers in a Three-Population Community with Trophic Relations

5.3.1. *Community of Three Trophic Levels*

In the model of a community of three populations interacting as a producer–consumer–predator system (Sec. 5.2.2; system (5.2.5)) the producer and the predator populations show unlimited growth for some values of the parameters.

To avoid this, let us modify system (5.2.5) by taking into account intraspecies competition in the producer population:

$$\begin{aligned}\dot{x} &= ax - b_1xy - ex^2\,,\\ \dot{y} &= -c_1y + d_1xy - b_2yz\,,\\ \dot{z} &= -c_2z + d_2yz\,.\end{aligned}\tag{5.3.1}$$

T.I. Eman (1966) studied this system for a trophic chain consisting of an arbitrary number of species with competition at the lowest trophic level. For reasons of a more complete presentation, we construct the bifurcation diagram and the phase portraits of system (5.3.1) and give their ecological interpretation. Setting $t = \tau/a$, $x = (c_1/d_1)u$, $y = (a/b_1)v$, and $z = (c_1/b_2)w$ changes system (5.3.1) to

$$\begin{aligned}\dot{u} &= u(1 - v - \epsilon u)\,,\\ \dot{v} &= -\gamma_1 v(1 - u + w)\,,\\ \dot{w} &= -\gamma_2 w(\alpha - v)\,,\end{aligned}\tag{5.3.2}$$

where the scaled parameters are expressed in terms of the initial parameters as for system (5.2.6), and $\epsilon = ec_1/ad_1$.

The equilibria of the system are

$$\mathbf{O}\ (u = v = w = 0)\,;$$

$$\mathbf{A}_1\ (u = 1/\epsilon,\ v = w = 0)\,;$$

$$\mathbf{A}_2\ (u = 1,\ v = 1 - \epsilon,\ w = 0)\,;$$

$$\mathbf{A}_3\ \left(u = \frac{1}{\epsilon}(1 - \alpha),\ v = \alpha,\ w = \frac{1}{\epsilon}(1 - \alpha) - 1\right)\,.$$

There are two curves of saddle-node bifurcations in the (α, ϵ)-plane of the equilibria A_1 and A_2, and A_2 and A_3, respectively, given by the equations $a_{12} : \epsilon = 1$ and $a_{23} : \alpha = 1 - \epsilon$ (Fig. 5.3.1). The nontrivial equilibrium A_3, corresponding to the coexistence of all three species, is in the positive octant of the phase space only for parameters from region *3*, that is, for $\alpha < 1 - \epsilon$. The equilibrium A_3 is always stable in the first octant. The parameter curves a_{12} and a_{23} complete the bifurcation diagram of the system. The corresponding phase portraits are shown in Fig. 5.3.1*b*.

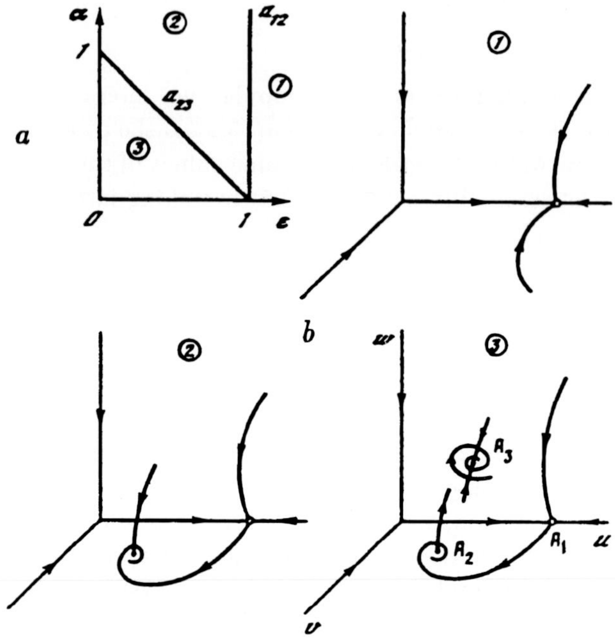

Fig. 5.3.1. Bifurcation diagram (*a*) and phase portraits (*b*) of system (5.3.2).

Ecologically, if $\epsilon > 1$ the capacity of the producer's ecological niche is small. Correspondingly, the stationary size of the producer population, determined by the resources available to it, is so small that it is unable to feed the consumer population. The isolated producer population is stable against the consumer's invasion.

If $1 > \epsilon > 1 - \alpha$ the niche capacity of the producer population is big enough to ensure the existence of the consumer population. Nevertheless, the stationary size of the consumer population is so small that it cannot feed the predator. The equilibrium A_2, corresponding to the coexistence of producer and consumer, is stable against the predator's invasion.

And finally, if $\epsilon < 1 - \alpha$ the niche capacity of the producer population is so large that the consumer population consuming the producer is now big enough for the predator to exist. The coexistence of all three species is stable.

5.3.2. *Two-Predators–One-Prey*

We showed that the systems of differential equations describing the dynamics of a two-predators–one-prey and that of a one-predator–two-preys community are formally equivalent (by means of reversing the direction of time). The situation changes when we allow for competition. The system describing the dynamics of a two-predators–one-prey community by taking into account competition among the prey takes the form

$$\begin{aligned}\dot{x} &= ax - b_1 x y_1 - b_2 x y_2 - e x^2\,,\\ \dot{y}_1 &= -c_1 y_1 + d_1 x y_1\,,\\ \dot{y}_2 &= -c_2 y_2 + d_2 x y_2\,.\end{aligned} \tag{5.3.3}$$

Setting $t = \tau/a$, $x = (a/e)u$, $y_1 = a/b_1$, and $y_2 = (a/b_2)v_2$ changes (5.3.3) to

$$\begin{aligned}\dot{u} &= u(1 - v_1 - v_2 - u)\,,\\ \dot{v}_1 &= -\gamma_1 v_1(\alpha - u)\,,\\ \dot{v}_2 &= -\gamma_2 v_2(\beta - u)\,,\end{aligned} \tag{5.3.4}$$

where $\gamma_1 = d_1/e$, $\gamma_2 = d_2/e$, $\alpha = c_1 e/a d_1$, and $\beta = c_2 e/a d_2$.

The equilibria of the system are

$$\begin{aligned}&\mathbf{O}\ (u = v_1 = v_2 = 0)\,;\\ &\mathbf{A}_1\ (u = 1,\ v_1 = v_2 = 0)\,;\\ &\mathbf{A}_2\ (u = \alpha,\ v_1 = 1 - \alpha,\ v_2 = 0)\,;\\ &\mathbf{A}_3\ (u = \beta,\ v_1 = 0,\ v_2 = 1 - \beta)\,.\end{aligned}$$

The points A_2 and A_3 are in the first quadrants of their respective coordinate planes for $\alpha < 1$ and $\beta < 1$, respectively. They are always stable nodes or foci in their coordinate planes.

Thus, the behavior of system (5.3.4) is independent of the parameters γ_1 and γ_2. The bifurcation diagram in the (α, β)-plane and the corresponding phase portraits are shown in Fig. 5.3.2. The phase portraits for regions *1* and *2* coincide with those shown in Fig. 5.3.1. For $\alpha = \beta < 1$ the phase portrait exhibits a straight line of non-isolated equilibria.

It can be seen from Fig. 5.3.2 that, in the general, case the system has one globally stable equilibrium. Depending on the parameters, it corresponds either to the extinction of both predators (when the producer's niche capacity does not provide a stationary size of the producer population large enough to maintain even one of the predators) or to the extinction of the predator that uses the prey resources in a less efficient way.

The two-predators–one-prey system with competing prey and predator saturation has been extensively studied (Hsu, 1978; Hsu *et al.*, 1978; Koch, 1974; Smith, 1982; Kirlinger, 1988). It has been shown that for certain values of the parameters the saturation of predators may make a coexistence of the predators possible when the same prey is consumed, but only in an oscillatory regime.

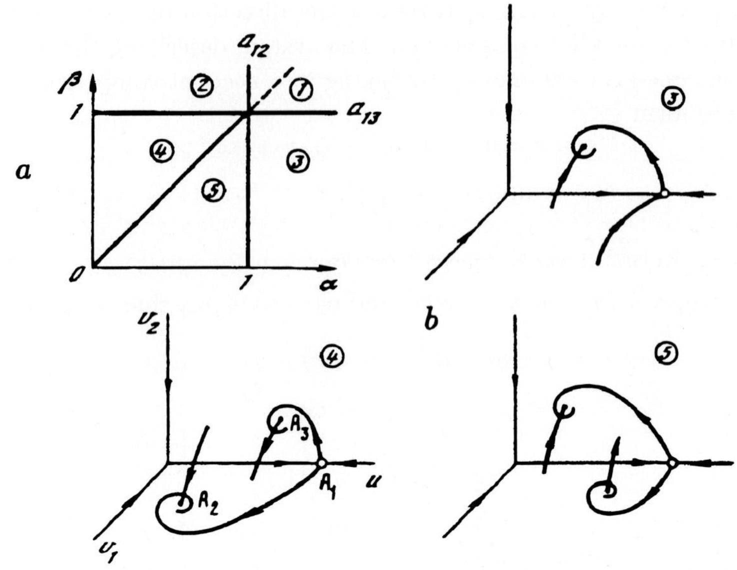

Fig. 5.3.2. Bifurcation diagram (a) and a set of phase portraits (b) of system (5.3.4).

5.3.3. *Trophic Cell*

Consider the dynamics of a trophic network's elementary cell by taking into account intraspecies competition in the producer population. A community of three such populations can be described by

$$\begin{aligned} \dot{x} &= ax - b_1xy - b_3xz - ex^2\,, \\ \dot{y} &= -c_1u + d_1xy - b_2yz\,, \\ \dot{z} &= -c_2z + d_2yz + d_3xz\,. \end{aligned} \tag{5.3.5}$$

Setting $t = \tau/a$, $x = (c_1/d_1)u$, $y = (a/b_1)v$, and $z = (c_1/b_2)w$ changes this system to

$$\begin{aligned} \dot{u} &= u(1 - v - \beta w - \epsilon u)\,, \\ \dot{v} &= -\gamma_1 v(1 - u + w)\,, \\ \dot{w} &= -\gamma_2 w(\alpha - v - \delta u)\,, \end{aligned} \tag{5.3.6}$$

where $\epsilon = ec_1/ad_1$, and the remaining parameters are expressed in terms of the initial parameters as for system (5.2.8).

The system has five equilibria:

$$\mathbf{O}\ (u = v = w = 0)\,;$$

$$\mathbf{B}\ (u = 1/\epsilon,\ v = w = 0)\,;$$

$$\mathbf{A}_1\ (u = 1,\ v = 1 - \epsilon,\ w = 0)\,;$$

$$\mathbf{A}_2 \quad \left(u = \frac{\alpha}{\delta},\ v = 0,\ w = \frac{1}{\beta}(1 - \epsilon\alpha/\delta)\right);$$

$$\mathbf{A}_3 \quad \left(u = \frac{1+\beta-\alpha}{\beta-\delta+\epsilon},\ v = \frac{\alpha(\beta+\epsilon)+\delta(\beta+1)}{\beta-\delta+\epsilon},\ w = \frac{1-\alpha+\delta-\epsilon}{\beta-\delta+\epsilon}\right).$$

The structure of the complete bifurcation diagram in $(\alpha, \beta, \delta, \epsilon)$-space can be imagined by considering its cross sections with (α, δ)-planes for arbitrary values of β. Those cross section depends on the sign of $\epsilon - 1$.

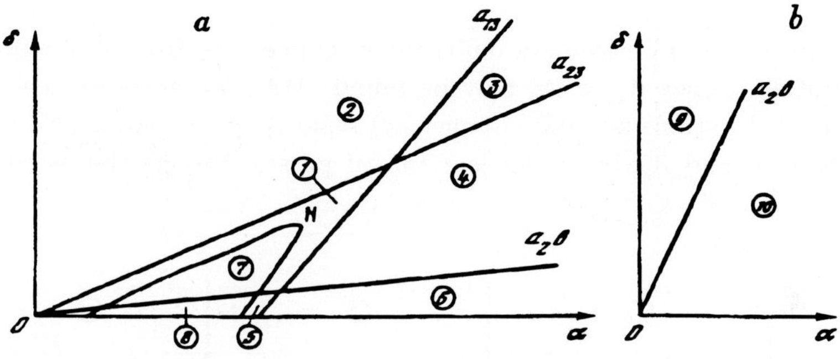

Fig. 5.3.3. The bifurcation diagram in the (α, δ)-plane of system (5.3.6) for $\epsilon < 1$ (a) and for $\epsilon > 1$ (b).

We start with the bifurcation diagram in the (α, δ)-plane for small values of $\epsilon \ll 1$ (Fig. 5.3.3*a*). This bifurcation diagram, like that of system (5.2.8) describing the dynamics of the trophic cell in the absence of competition among the producers, shows saddle-node bifurcation curves corresponding to bifurcations of A_3 with A_1 and A_2, respectively. Their equations are given by $a_{13} : \delta = \alpha + \epsilon - 1$ and $a_{23} : \delta = \alpha(\beta+\epsilon)/(\beta+1)$. There is one more curve, namely the saddle-node bifurcation curve where A_2 meets B, with the equation $a_2b : \delta = \epsilon\alpha$. These are not all bifurcation curves in the bifurcation diagram of the system. It contains an Andronov-Hopf bifurcation curve N, where A_3 loses its stability and a small stable limit cycle is born. The equation of the Andronov-Hopf curve is too bulky to be given here, nevertheless, it is obvious from continuity considerations that the Andronov-Hopf curve is stuck in the cone formed by the bifurcation curves a_{13} and a_{23} for $\epsilon \to 0$. A numerical calculation of the Andronov-Hopf curve done by E.A. Aponina for $\epsilon = 0.01$ and for $\epsilon = 0.1$ has confirmed the correctness of the bifurcation diagram in Fig. 5.3.3*a*.

For $\epsilon > 1$ the bifurcation diagram has the shape shown in Fig. 5.3.3*b*. Recall that only those curves are plotted that correspond to bifurcations in the positive octant.

The complete set of phase portraits of the system is shown in Fig. 5.3.4 for the corresponding parameter regions.

We can see that the bifurcation diagram of system (5.3.6) is quite complicated and shows an extensive collection of dynamic regimes. Depending on the parameters, there can be either one or two attractors in the phase space. If there is a single attractor then it is one of the following objects:

- the point B (region *10*),
- the point A_1 (regions *4* and *6*),
- the point A_2 (regions *2* and *9*),
- the point A_3 (regions *1* and *5*), and
- a three-dimensional stable limit cycle (regions *7* and *8*).

For parameters from region *3* a multistable regime is realized, depending on the initial condition, either A_1 or A_2 may be found. We have not specifically studied the location of the stable manifold (a surface) separating the basins of attraction of the equilibria A_1 and A_2, but it is clear that it passes through the w-axis and the point A_3.

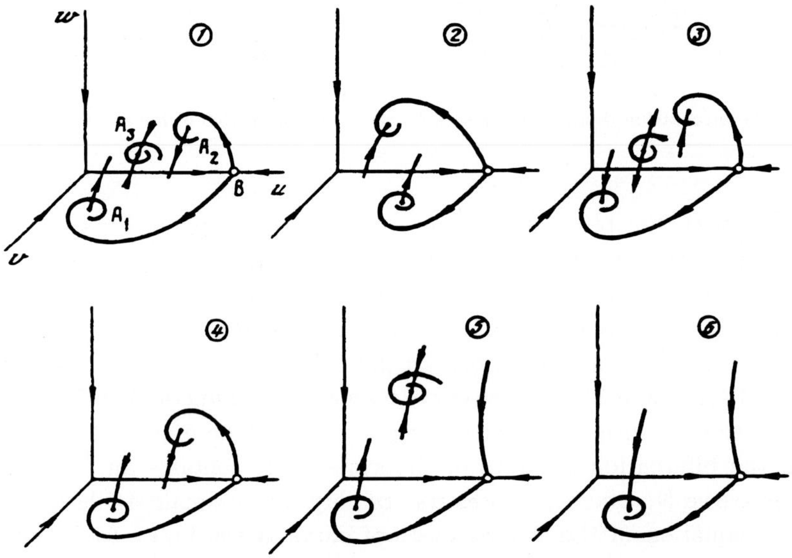

Fig. 5.3.4. Phase portraits of system (5.3.6).

Let us now interpret these results from the ecological point of view. The meaning of the limitations determining the boundary of parameter region *10*, and the behavior of the system for parameters from that region are obvious: For $\epsilon > \max\{1, \alpha/\delta\}$ the producer's ecological niche has such a small capacity that the producer population can neither ensure the existence of the consumer ($\epsilon > 1$) nor that of the predator in the absence of the consumer ($\epsilon > \alpha/\delta$). The situation for region *9* is very similar: For $\alpha/\delta > \epsilon > 1$ the capacity of the producer's ecological niche and the rates at which the consumer and the predator consume the producer are not enough for the producer population to feed the consumer, but a stationary existence of the predator population in the absence of the consumer can be maintained. Under these circumstances the predator should no longer be called a predator, because it completely switches to herbivorous feeding.

The inequality $1 > \epsilon > \alpha/\delta$ determining the boundary of regions *5*, *6* and *8*, describes a situation when the producer's niche capacity suffices, so that the producer is able to feed the consumer, but is not enough for the predator to exist in the absence of the consumer.

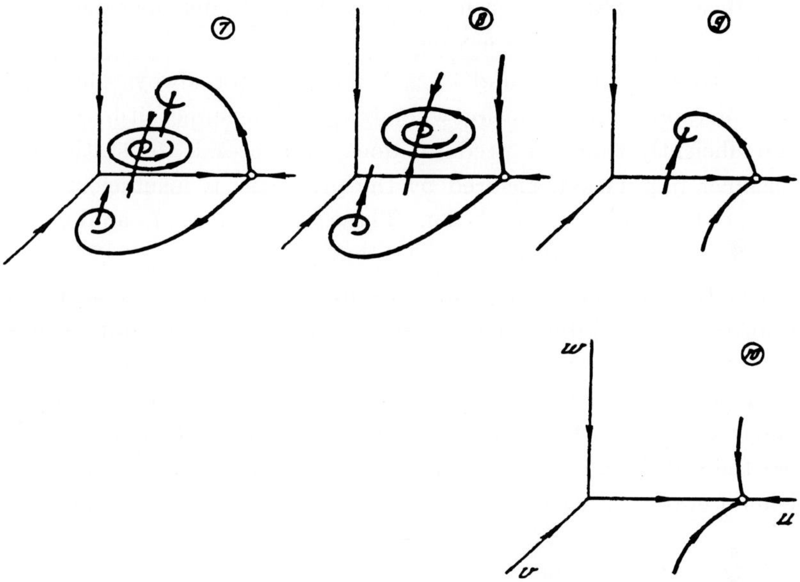

Fig. 5.3.4. (Continued)

Here two things are possible: either (region *6*) the stationary size of the consumer population is insufficient to ensure the existence of the predator (and the latter is doomed to die out) or (regions *5* and *8*) the producer population can feed the predator population. In the latter case the stable coexistence of three populations is stationary for parameters from region *5* and oscillatory for parameters from region *8*.

Let us now consider the condition ($\epsilon < \min\{1, \alpha/\delta\}$) when the niche capacity of the producer and the rate of its being jointly consumed by the consumer and the predator is such as to make it possible for the producer population to separately feed each, the consumer and the predator population. Note that here the consumer and predator populations can be regarded not only as populations of neighboring trophic levels but also as a system interacting populations while consuming a common resource, namely the producer.

If we exclude the trophic interaction between the consumer and the predator then the system changes into the two-predators–one-prey system that was analysed in the previous section. Depending on the parameters, one of the two species consuming the producer drives out the other completely for any initial conditions.

What changes or stays the same in the two-consumer–one-producer system when we introduce trophic relations between the consumers? (This means changing the two-consumer–one-producer system into a producer–consumer–predator system with facultative herbivorous food habits of the predator.)

It can be seen in Fig. 5.3.4 that for $\delta > \alpha + \epsilon$ the predator always drives the producer to extinction. In other words, if the predator consumes the producer much more efficiently than the consumer does, then the consumer dies out, and the predator switches to herbivorous feeding.

For $\delta < \alpha(\beta + \epsilon)/(\beta + 1)$ and $\alpha > 1$ the predator always dies out. These inequalities correspond to a situation when first, the consumer utilizes the producer much more efficiently than the predator does, and second, the stationary density of the consumer population, ensured by the producer, is insufficient to maintain the existence of the predator population. These two regimes are analogous to those observed in the two-predators–one-prey system.

Apart from the above, there are some intermediate values of δ/α which are shown as a cone formed by the bifurcation curves a_{13} and a_{23} in the bifurcation diagram. For parameters from this cone the system can exhibit two types of dynamic regimes. In the left-hand lower corner of the cone there can be either one nontrivial stable equilibrium (region *1*) or a three-dimensional stable limit cycle (region *7*), depending on whether the stable coexistence of the three species is stationary or oscillatory. In the right-hand upper corner of the cone (region *3*) a multistable regime is realized in the system, so that for fixed parameters either the predator or the consumer dies out, depending on their initial population sizes.

The elementary cell of trophic networks we have studied is probably the simplest ecological model in which, depending on the parameters, the three species can coexist (both in a stationary and oscillatory regime) or exist in a multistable mode characterized by the extinction of one population (depending on the initial values).

Earlier numerical experiments showed that oscillations are possible in a Volterra system describing the interaction of six species located on three trophic levels (Garfinkel, Richard, 1964; Maynard Smith, 1974). This possibility was more thoroughly investigated for ecological models describing a closed cycle of substances in systems of four or six species (Bazykin, 1982).

Recall that in the predator–prey community, described by a system of two differential equations, using bilinear interaction between the predator and the prey is insufficient to cause stable oscillations, so that some additional destabilizing factors are required. However, oscillations in the elementary cell of the trophic network, and naturally in more complicated communities, can arise for certain values of the parameters, even when bilinear interaction between species is used.

5.3.4. *One-Predator–Two-Preys*

Let us now consider qualitatively what new effects are caused by interspecies competition among the prey populations in a one-predator–two-preys community.

The population dynamics is described by the following system of differential equations

$$\begin{aligned} \dot{x}_1 &= a_1x_1 - b_1x_1y - e_{11}x_1^2 - e_{12}x_1x_2\,, \\ \dot{x}_2 &= a_2x_2 - b_2x_2y - e_{21}x_1x_2 - e_{22}x_2^2\,, \\ \dot{y} &= -cy + d_1x_1y + d_2x_2y\,. \end{aligned} \tag{5.3.7}$$

Setting $t = \tau/c$, $x_1 = (c/e_{11})u_3$, $x_2 = (c/e_{22})u_2$, and $y = v$ changes (5.3.7) to

$$\begin{aligned} \dot{u}_1 &= u_1(\alpha_1 - \beta_1 v - u_1 - \epsilon_1 u_2)\,, \\ \dot{u}_2 &= u_2(\alpha_2 - \beta_2 v - u_2 - \epsilon_2 u_1)\,, \\ \dot{v} &= -v(1 - \delta_1 u_1 - \delta_2 u_2)\,, \end{aligned} \tag{53.8}$$

where $\alpha_{1,2} = a_{1,2}/c$, $\beta_{1,2} = b_{1,2}/c$, $\epsilon_1 = e_{12}/e_{22}$, $\epsilon_2 = e_{21}/e_{11}$, $\delta_1 = d_1/e_{11}$, and $\delta_2 = d_2/e_{22}$.

One of the parameters β_1 or β_2 can be removed by scaling y, but we do not do this in order to keep a symmetric notation.

System (5.3.7) was first studied by Parish and Saila (1970), who showed that for some values of the parameters the presence of a predator in a community can ensure the coexistence of competing prey populations (which is impossible in the absence of a predator). There are experimental findings (see, e.g., Paine, 1966) which illustrate this theoretical conclusion and seem to corroborate the adequacy of model (5.3.7). The same qualitative effect was later studied for some special (equality-type) requirements imposed on the parameters (Cramer, May, 1972; Hsu, 1981). Fujii (1979) ascertained that only one stable limit cycle can exist in this system within a certain range of parameters.

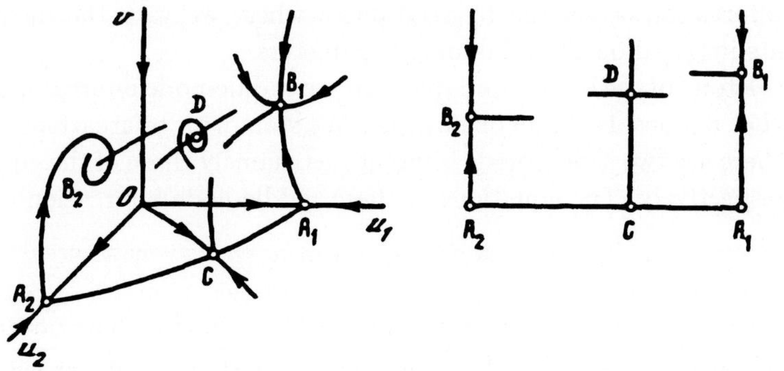

Fig. 5.3.5. Location and notation of equilibria in the phase portrait of system (5.3.8) and in its schematic representation.

System (5.3.7) augmented by a term describing competition among predators was numerically studied for different parameter values. Vance (1978) detected the existence of complicated periodic regimes and plotted some bifurcation curves in the $(\alpha_1, \epsilon_1 - \epsilon_2)$plane (in our notation) for fixed values of the other parameters. Gilpin (1979) discovered and studied numerically a phase portrait with a chaotic regime, which he called "spiral chaos", using Rössler's terminology (Rössler, 1976). Arneodo *et al.* (1980) investigated how system (5.3.7) depends on the parameter α_2 (in our notation) for fixed values of the other parameters. The authors discovered complicated dynamic behavior and suggested an explanation of its bifurcation structure. Unfortunately, the signs chosen by these authors for the parameters do not permit any ecological interpretation.

Below we try to study as completely as possible the behavior of system (5.3.8) and how it depends on all parameters. There are seven equilibria of this system (Fig. 5.3.5):

$$\begin{aligned}
&\mathbf{O}\ (u_1 = u_2 = v = 0);\\
&\mathbf{A}_1\ (u_1 = \alpha_1,\ u_2 = v = 0);\\
&\mathbf{A}_2\ (u_1 = 0,\ u_2 = \alpha_2,\ v = 0);\\
&\mathbf{B}_1\ \left(u_1 = 1/\delta_1,\ u_2 = 0,\ v = \frac{\alpha_1\delta_1 - 1}{\beta_1\delta_1}\right);\\
&\mathbf{B}_2\ \left(u_1 = 0,\ u_2 = 1/\delta_2,\ v = \frac{\alpha_2\delta_2 - 1}{\beta_2\delta_2}\right);\\
&\mathbf{C}\ \left(u_1 = \frac{\alpha_1 - \alpha_2\epsilon_1}{1 - \epsilon_1\epsilon_2},\ u_2 = \frac{\alpha_2 - \alpha_1\epsilon_2}{1 - \epsilon_1\epsilon_2},\ v = 0\right);\\
&\mathbf{D}\ \text{(coordinates can be determined from the}\\
&\qquad \text{corresponding system of algebraic equations)}
\end{aligned} \tag{5.3.9}$$

System (5.3.8) depends on eight parameters. We begin our study by considering the bifurcation set in the (α_1, α_2)-plane where we vary the inequality-type requirements imposed on the remaining parameters.

The "skeleton" of the bifurcation diagram are saddle-node bifurcations of various equilibria. Let us consider them one by one. We start with bifurcations in the plane $\{v = 0\}$. There are two saddle-node bifurcations, namely those of the equilibria $\mathsf{A}_1\mathsf{C}$ and $\mathsf{A}_2\mathsf{C}$, respectively. The equations of these saddle-node bifurcation curves are

$$\begin{aligned}
\mathbf{A}_1\mathbf{C} &: \alpha_2 = \epsilon_2\alpha_1\,,\\
\mathbf{A}_2\mathbf{C} &: \alpha_1 = \epsilon_1\alpha_2\,.
\end{aligned} \tag{5.3.10}$$

This means that in the (α_1, α_2)-plane the bifurcation curves pass through the origin with a positive slope. They can have two relative positions depending on the sign of $\epsilon_1\epsilon_2 - 1$. We consider first the case $\epsilon_1\epsilon_2 < 1$ (Fig. 5.3.6).

Here, and throughout this section, we use numbers without indices for phase portraits that do not change qualitatively by exchanging u_2 and u_1. On the other hand, we use numbers with the indices 1 and 2 for those phase portraits that to change if u_1 and u_2 are exchanged.

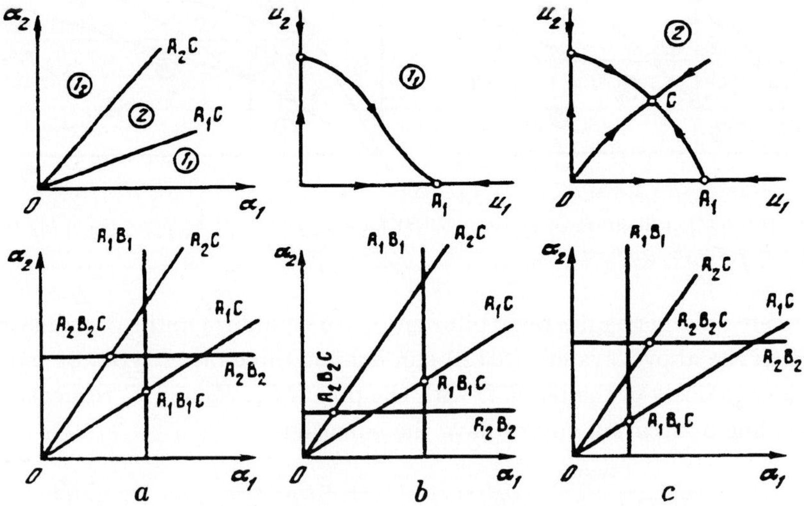

Fig. 5.3.6. Relative positions of bifurcation curves in the bifurcation diagram of system (5.3.8) for $\epsilon_1\epsilon_2 < 1$ and phase portraits in the plane $\{v = 0\}$ for regions 1_1 and 2.

Let us now consider the bifurcations that occur in the coordinate planes $\{u_2 = 0\}$ and $\{u_1 = 0\}$. These are saddle-node bifurcations of the equilibria $\mathsf{A}_1\mathsf{B}_1$ and $\mathsf{A}_2\mathsf{B}_2$, respectively. The corresponding bifurcation curves are given by the equations

$$\begin{aligned} \mathbf{A}_1\mathbf{B}_1 &: \alpha_1 = 1/\delta_1 \,, \\ \mathbf{A}_2\mathbf{B}_2 &: \alpha_2 = 1/\delta_2 \,, \end{aligned} \tag{5.3.11}$$

which are parallel to the coordinate axes.

The curves $\mathsf{A}_1\mathsf{B}_1$ and $\mathsf{A}_2\mathsf{B}_2$ can have the different relative positions with the rays $\mathsf{A}_1\mathsf{C}$ and $\mathsf{A}_2\mathsf{C}$ according to (Fig. 5.3.6):

(a) $\epsilon_1\delta_1 < \delta_2$, $\epsilon_2\delta_2 < \delta_1$;
(b) $\epsilon_1\delta_1 > \delta_2$, $\epsilon_2\delta_2 < \delta_1$;
(c) $\epsilon_1\delta_1 < \delta_2$, $\epsilon_2\delta_2 > \delta_1$.

The phase portraits b and c, shown in Fig. 5.3.6, are equivalent in the sense that they are exchanged by exchange the respective indices. Therefore, it is sufficient for a complete study to consider only one type of them.

The points where the curves $\mathsf{A}_1\mathsf{B}_1$ and $\mathsf{A}_1\mathsf{C}$, as well as the curves $\mathsf{A}_2\mathsf{B}_2$ and $\mathsf{A}_2\mathsf{C}$, intersect corresponds to a bifurcations of conditional codimension two, namely simultaneous saddle-node bifurcations at A_1B_1C, and A_2B_2C.

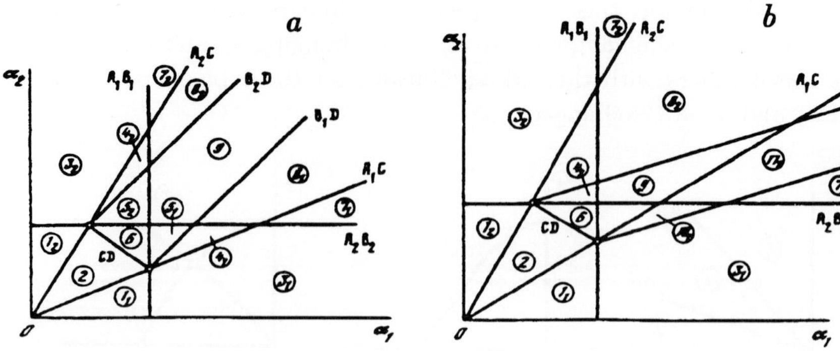

Fig. 5.3.7. Bifurcation diagram of system (5.3.8) for $\epsilon_1\epsilon_2 < 1, \epsilon_1 < \delta_2/\delta_1$, $\epsilon_2 < \delta_1/\delta_2$ (*a*) and for $1/\epsilon_1 > \beta_2/\beta_1 > \epsilon_2$, $\beta_2/\beta_1 < \epsilon_2$ (*b*).

There are no other sadlle-node bifurcations of equilibria in the coordinate planes. Apart from the above, system (5.3.8) can exhibit three other types of saddel-node bifurcations, namely of the point D with the points B_1, B_2 and C, respectively. The corresponding bifurcation curves have the equations

$$\begin{aligned}
\mathbf{B_1D}: \quad \alpha_2 &= \frac{\beta_2(\alpha_1\delta_1 - 1) + \epsilon_2\beta_1}{\beta_1\delta_1}, \\
\mathbf{B_2D}: \quad \alpha_1 &= \frac{\beta_1(\alpha_2\delta_2 - 1) + \epsilon_1\beta_2}{\beta_2\delta_2}, \qquad (5.3.12) \\
\mathbf{CD}: \quad \alpha_1(\delta_1 - \delta_2\epsilon_2) &+ \alpha_2(\delta_2 - \delta_1\epsilon_1) = 1 - \epsilon_1\epsilon_2 .
\end{aligned}$$

It can easily be seen that the straight line CD passes through the two points $\mathsf{A_1B_1C}$ and $\mathsf{A_2B_2C}$, while the straight lines $\mathsf{B_1D}$ and $\mathsf{B_2D}$ pass through the points $\mathsf{A_1B_1C}$ and $\mathsf{A_2B_2C}$, respectively, having the identical positive slope β_2/β_1.

Note that the points $\mathsf{A_1B_1C}$ and $\mathsf{A_2B_2C}$, which have conditional codimension two as they are at the intersection of two bifurcation curves, correspond to the simultaneous merging of the four equilibria A_1, B_1, C and D and that of A_2, B_2, C and D, respectively.

Figures 5.3.7 and 5.3.8 show all possible relative positions of bifurcation curves for the case $\epsilon_1\epsilon_2 < 1$. The corresponding phase portraits are schematically shown in Fig. 5.3.9 (see also Fig. 5.3.5).

Note that the points B_1 and B_2 are always stable in their repective coordinate planes and, depending on the parameter values, may be nodes or foci. As will become clear in the further discussion, the general dynamic regime of the system turns out to be significantly different in these two cases. Let us consider the case when one of the points B_1 and B_2, or both of them, are foci until the end of this section. We start with investigating the neighborhoods in parameter space of the bifurcation curves $\mathsf{A_1B_1}$ and $\mathsf{A_2B_2}$, that is, we take the parameters such that B_1

and B_2 are stable nodes. In this case, the three-dimensional phase portraits of the system may be adequately presented by means of their two-dimensional projections onto the plane $\{u_1/\alpha_1 + u_2/\alpha_2 = 1\}$. The bifurcation diagrams show only the parts of the bifurcation curves corresponding to the bifurcations in the first octant of the phase space. Note that for $\epsilon_1\epsilon_2 < 1$ the saddle-node bifurcation curves completely describe the bifurcation diagram of the system.

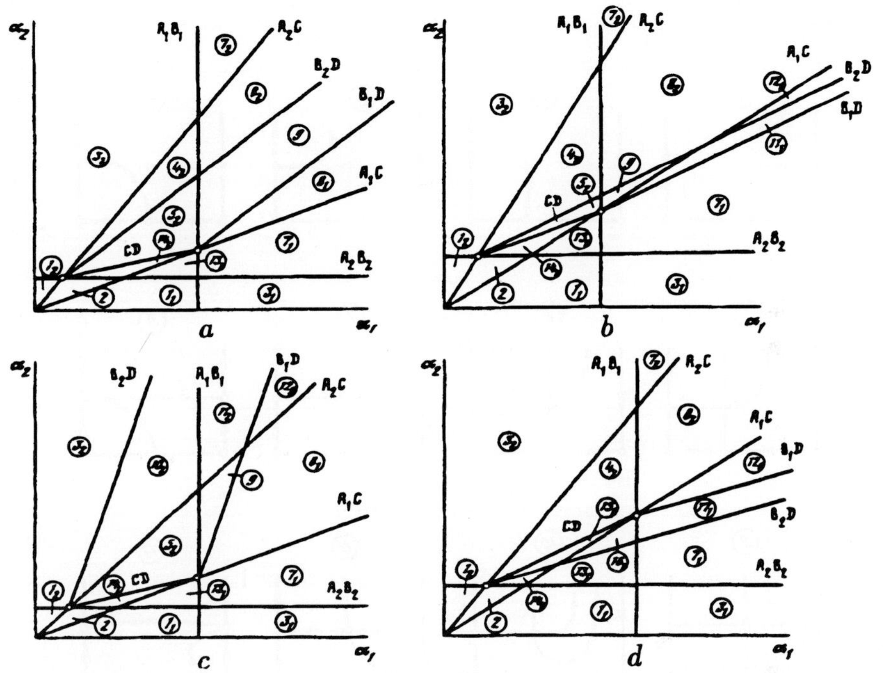

Fig. 5.3.8. Bifurcation diagram of system (5.3.8) for $\epsilon_1\epsilon_2 < 1$, $\epsilon_1 > \delta_2/\delta_1$, $\epsilon_2 < \delta_1/\delta_2$, with the additional condition that $1/\epsilon_1 > \beta_2/\beta_1 > \epsilon_2$ (*a*), that $\beta_2/\beta_1 > 1/\epsilon_1$ (*b*), that $\epsilon_2 > \beta_2/\beta_1 > -\frac{\delta_1-\delta_2\epsilon_2}{\delta_2-\delta_1\epsilon_1}$ (*c*), and that $\beta_2/\beta_1 < -\frac{\delta_1-\delta_2\epsilon_2}{\delta_2-\delta_1\epsilon_1}$ (*d*).

Before turning to the next case ($\epsilon_1\epsilon_2 > 1$), let us give an ecological interpretation of the above results. In the first place, we should note that the inequality $\epsilon_1\epsilon_2 < 1$ suggests that the interspecies competition among the two prey populations is weaker compared to the intraspecies competition, and, therefore, both prey populations can stably coexist in the absence of a predator. When the predator is introduced into the community, the following regimes are possible (see the phase portraits):

1. One globally stable equilibrium:

(a) one prey population without the predator (*1, 13*);
(b) coexistence of prey populations without the predator (*2, 14*);
(c) one prey population with the predator (*3, 4, 7, 8, 12*);
(d) coexistence of all three species (*5, 6, 9, 10, 11*).

2. Multistable regimes:

(a) either one of the prey populations coexists with the predator, or the other exists without the predator (*16*);

(b) each of the prey populations coexists with the predator, but not simultaneously (*17*);

(c) either one of the prey populations coexists with the predator or both of them coexist without the predator (*15*).

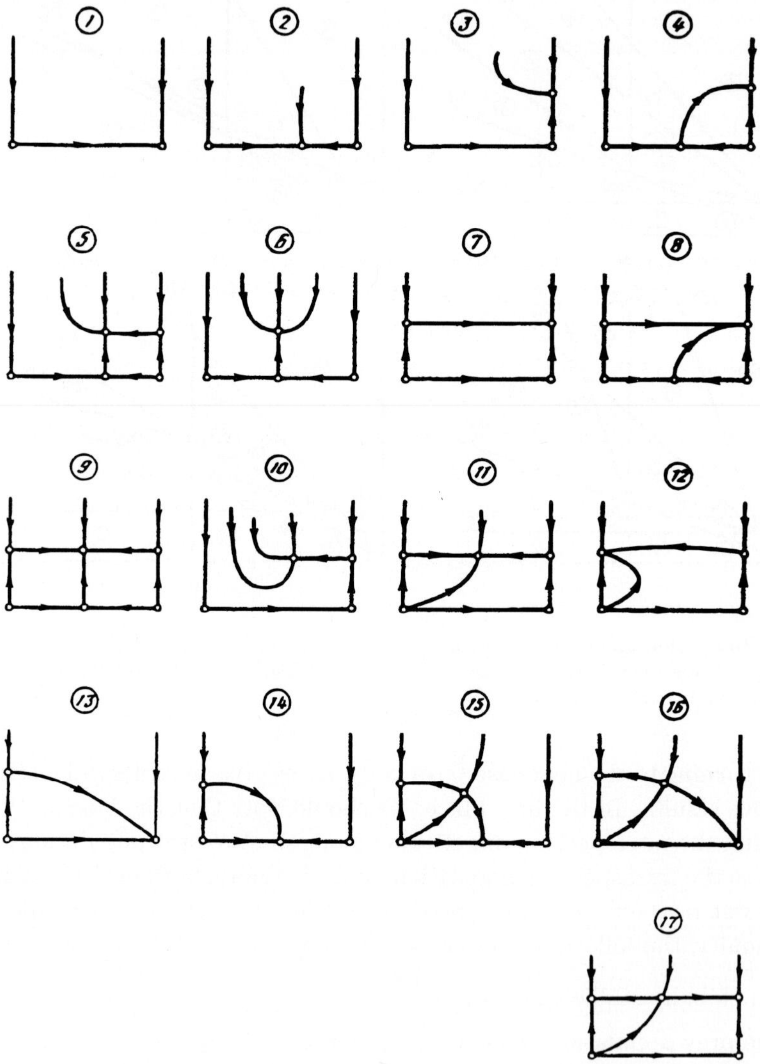

Fig. 5.3.9. Phase portraits of system (5.3.8) for $\epsilon_1\epsilon_1 < 1$. Phase portraits *2, 6* and *9* correspond to parameter regions without index, and the remaining regions have index 1.

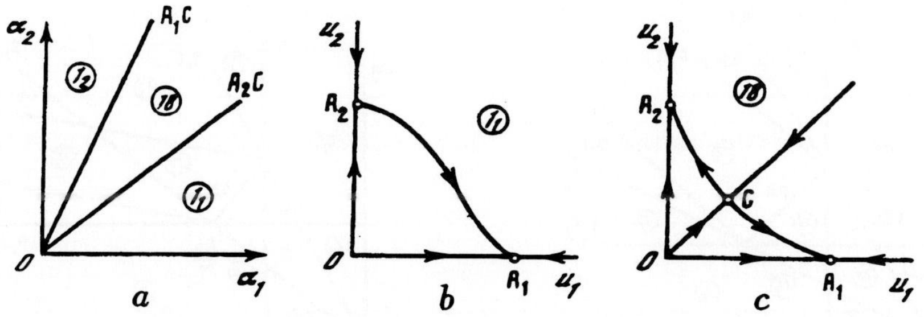

Fig. 5.3.10. Relative positions of the bifurcation curves $\mathsf{A}_1\mathsf{C}$ and $\mathsf{A}_2\mathsf{C}$ in the (α_1, α_2)-plane for $\epsilon_1\epsilon_2 > 1$ (a), and the corresponding phase portraits in the plane $\{v = 0\}$ (b and c). (The numbering of phase portraits of system (3.5.8), started for the case $\epsilon_1\epsilon_2 < 1$, continues with *18*.)

Note that the four types of globally attracting regimes we found in the system are no surprise, while it is of great interest that also multistable regimes (where either one of the prey populations outcompetes the other, or both of them coexisted in a stable manner) appears when the predator is introduced into the community.

Let us now turn to the case when $\epsilon_1\epsilon_2 > 1$ (Fig. 5.3.10). The main difference from the previous case lies in the fact that, while the equilibrium C is a stable node in the plane $\{v = 0\}$ for $\epsilon_1\epsilon_2 < 1$, it changes into a saddle (see Fig. 5.3.10) for $\epsilon_1\epsilon_2 > 1$, and the prey populations cannot coexist in the absence of a predator.

The straight lines of saddle-node bifurcations are also given by equations (5.3.10–5.3.12). All their possible relative positions are shown in Figs. 5.3.11 and 5.3.12, and the corresponding phase portraits are presented in Fig. 5.3.13. It is important that in all except one of the cases the bifurcation diagrams are given completely by the saddle-node bifurcation curves. The only exception, and the most interesting case, is the bifurcation diagram shown in Fig. 5.3.12d, for which

$$\begin{aligned} \beta_2/\beta_1 &> -\frac{\delta_1 - \delta_2\epsilon_2}{\delta_2 - \delta_1\epsilon_1}\,, \\ \delta_2 &> \delta_1\epsilon_1\,, \\ \delta_1 &< \delta_2\epsilon_2\,. \end{aligned} \tag{5.3.13}$$

The situation remains the same if all three inequalities change signs, since this transformation correspond to a change of parameters. Graphically, the inequalities (5.3.13) mean that the pair of bifurcation curves $\mathsf{B}_1\mathsf{D}$ and $\mathsf{B}_2\mathsf{D}$ has greater slope than the curve CD.

In this case (Fig. 5.3.14), the points $\mathsf{A}_1\mathsf{B}_1\mathsf{C}$ and $\mathsf{A}_2\mathsf{B}_2\mathsf{C}$ of the bifurcation diagram are connected by the saddle-node curve of the equilibria C and D, and by, at least, two more bifurcation curves. They are an Andronov-Hopf curve N of the equilibrium D and a curve L of parameters for which there is a heteroclinic cycle, formed by the coordinate axes and the unstable manifold of B_2, which coincides with the stable manifold of C.

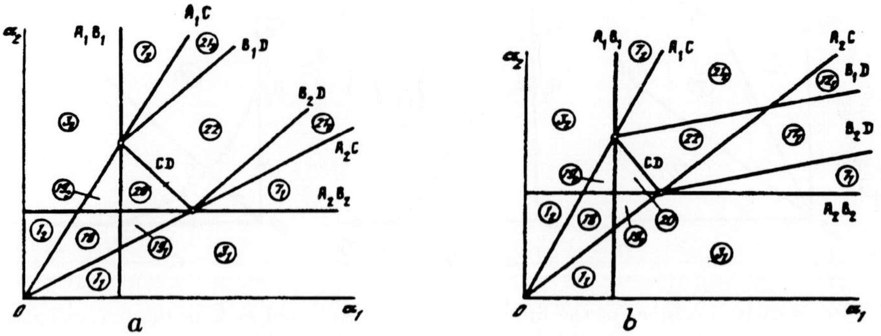

Fig. 5.3.11. Bifurcation diagram of system (5.3.8) in the (α_1, α_2)-plane for $\epsilon_1\epsilon_2 > 1, \epsilon_1 > \delta_2/\delta_1, \epsilon_2 > \delta_1/\delta_2$, with the additional condition that $\epsilon_2 > \beta_2/\beta_1 > 1/\epsilon_1$ (*a*) , and that $\beta_2/\beta_1 < 1/\epsilon_1$ (*b*).

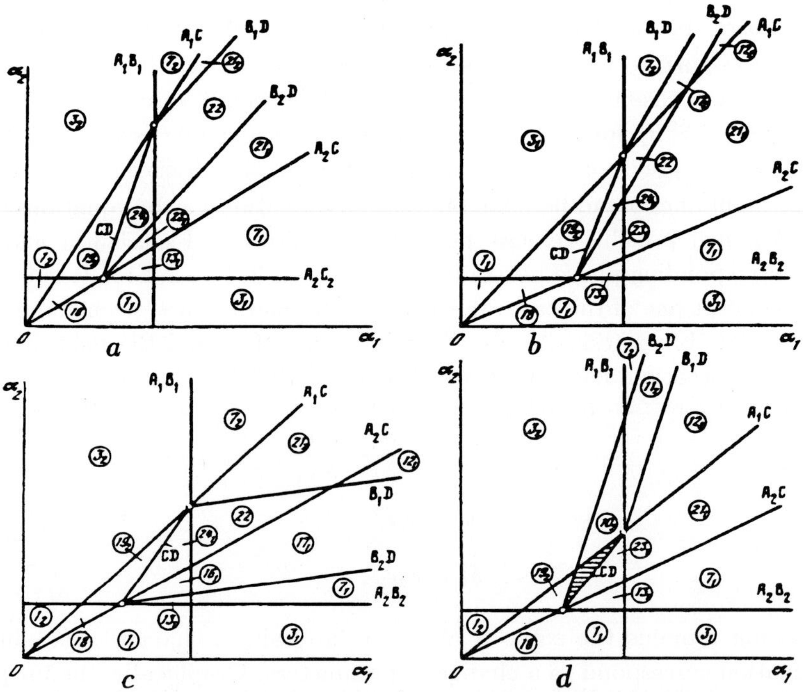

Fig. 5.3.12. Bifurcation diagram of system (5.3.8) in the (α_1, α_2)-plane for $\epsilon_1\epsilon_2 < 1, \epsilon_1 < \delta_2/\delta_1, \epsilon_2 > \delta_1/\delta_2$ with the additional condition that $\epsilon_2 > \beta_2/\beta_1 > 1/\epsilon_1$ (*a*), that $\beta_2/\beta_1 < 1/\epsilon_1$ (*b*), that $-\frac{\delta_1-\delta_2\epsilon_2}{\delta_2-\delta_1\epsilon_1} > \beta_2/\beta_1 > \epsilon_2$ (*c*), and that $\beta_2/\beta_1 > -\frac{\delta_1-\delta_2\epsilon_2}{\delta_2-\delta_1\epsilon_1}$ (*d*). The structure of the shaded region is shown in Fig. 5.3.14.

Note that the points A_1B_1C and A_2B_2C correspond to the merging of three equilibria in the phase portrait, that is, they are points of parameters for which two eigenvalues of the system vanish (on the condition that there is an additional

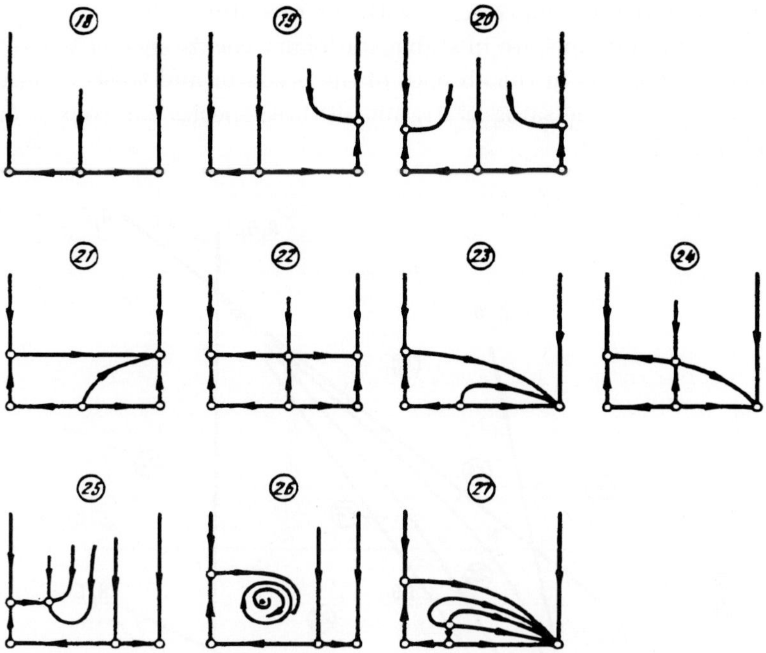

Fig. 5.3.13. Phase portraits of system (5.3.8) for $\epsilon_1\epsilon_2 > 1$. The numbering corresponds to that of the regions of Figs. 5.3.11 and 5.3.12. Phase portraits *18, 20* and *22* correspond to parameter regions without index, the remaining phase portraits have index 1.

degeneracy caused by the specific form of the system). Therefore, the fact that the Andronov-Hopf curve and the heteroclinic cycle curve arrive at these points represents the general case. For parameters not satisfying (5.3.13), those curves correspond to bifurcations that occur outside the positive octant (and they are not shown in the figures).

Let us now interpret the results from the ecological point of view. Recall that we consider the case ($\epsilon_1\epsilon_2 > 1$) when, in the absence of a predator, two competing prey populations cannot coexist. The introduction of a predator into the community can lead to the following.

1. Globally attracting regimes

(a) one prey population without the predator,
(b) one prey population with the predator, and
(c) stationary coexistence of all three populations.

2. Multistable regimes

(a) in the absence of the predator, either prey population exists,
(b) either one prey population coexists with the predator or the other exists without the predator,

(c) either prey population coexists with the predator,
(d) all three populations are in stable stationary coexistence or one of the prey populations exists in the absence of the predator and the competitor, and
(e) the situation is the same as d), but all three species can coexist only in an oscillatory regime.

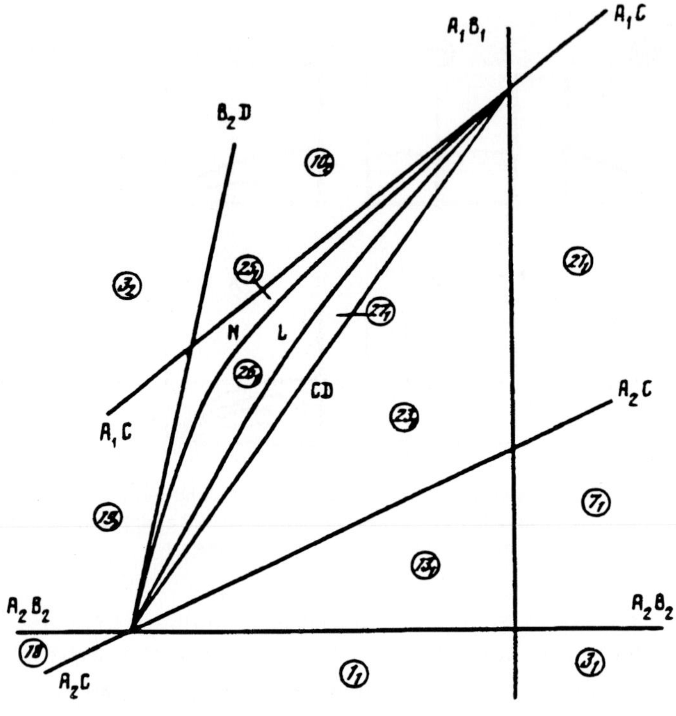

Fig. 5.3.14. Structure of the shaded parameter region in Fig. 5.3.12*d*.

In our opinion the most interesting results are the following:

1. The introduction of a predator into the community can ensure stable coexistence of competing prey populations, which is impossible in the absence of a predator.
2. The coexistence of all three species may either be globally stable or has a boundary of its basin of attraction in the phase space.
3. All three species may coexist in a stationary or an oscillatory regime in the absence of any special destabilizing factors.

We have mentioned above that it may be important for the general dynamics of the system whether the point B_2 in the plane $\{u_1 = 0\}$ is a stable node or a focus. For the majority of parameter regions this difference does not affect the qualitative behavior of the system. Nevertheless, this is important for parameters from the neighborhood of the boundary between regions *26* and *27*. Let us consider this situation more carefully.

The main results on the dynamics of a system of three differential equations for parameters close to which the phase portrait of the system exhibits a *saddle-node loop* or a *saddle-focus loop* are due to L. P. Shil'nikov (1963, 1965, 1970, 1980). The essence of his results is the following. The system can exhibit three different dynamic regimes, depending on, first, whether the equilibrium is a saddle-node or saddle-focus, and, second, on the sign of the saddle quantity σ. (Here $\sigma = |\lambda/\mu|$, where λ and μ are the largest negative and positive eigenvalues in the case of a saddle; in the case of saddle-focus λ is complex and in the above definition should be replaced by $\mathrm{Re}\lambda$.) Two of these three are analogous to the corresponding behavior of two-dimensional systems, whereas the third is qualitatively new.

1. $\sigma > 1$. Whatever the equilibrium (for both, a saddle-node or a saddle-focus), the homoclinic loop gives rise to a periodic motion, a stable limit cycle, when the parameter changes. If the parameter changes in the opposite direction the stable limit cycle breaks up as it becomes a homoclinic loop.
2. $\sigma < 1$, and the equilibrium is a saddle-node. Only a limit cycle of saddle type appears in the homoclinic loop bifurcation.
3. $\sigma < 1$, and the equilibrium is a saddle-focus. This is the most complicated case, and there exists an infinite (countable) number of closed trajectories in any small neighborhood of the homoclinic loop. (However, this does not exhaust all possible trajectories lying in the neighborhood of the homoclinic loop.) The bifurcation phenomena near such a point are also complicated, and they have not been completely studied in the general case. We shall return to some of them when we discuss the results of the numerical study. L. A. Belyakov (1981) studied the structure of a neighborhood of the following codimension-two bifurcation points: First, the intersection of the homoclinic loop curve and the "node-focus" curve (which is not a bifurcation curve) for $\sigma < 1$, and second, the point on the homoclinic loop curve where the saddle-quantity σ passes through one.

Shil'nikov's results can be easily applied to system (5.3.8) with the only difference that the heteroclinic cycle $\mathsf{A}_2\mathsf{B}_2\mathsf{C}\mathsf{A}_2$ appears instead of a homoclinic loop. The saddle-quantity is given by the expression

$$\sigma = \left| \frac{\lambda(\mathbf{A}_2)\lambda(\mathbf{C}) \cdot \max \mathrm{Re}\lambda_{1,2}(\mathbf{B}_2)}{\mu(\mathbf{A}_2)\mu(\mathbf{C})\mu(\mathbf{B}_2)} \right| , \qquad (5.3.14)$$

where λ and μ are the respectively negative and positive eigenvalues of the system at the corresponding points in the direction of motion along the heteroclinic cycle.

Let us examine the bifurcations occuring in a neighborhood of the heteroclinic cycle for a concrete numerical example. Together with E. A. Aponina and Yu. M. Aponin (Aponina, Aponin, Bazykin, 1982), the author studied the following one-predator–two-preys system:

$$\begin{aligned}\dot{u}_1 &= u_1(\alpha_1 - u_1 - 6u_2 - 4v)\,,\\ \dot{u}_2 &= u_2(\alpha_2 - u_2 - u_1 - 10v)\,,\\ \dot{v} &= -v(1 - 0.25u_1 - 4u_2 + v)\,.\end{aligned} \tag{5.3.15}$$

Unlike system (5.3.8), this system allows for the additional factor of intraspecies competition in the predator population. This factor results only in insignificant changes of the bifurcation diagram: the bifurcation curves $\mathsf{B}_1 D$ and $\mathsf{B}_2 D$ in the (α_1, α_2)-plane are no longer parallel, and the Andronov-Hopf curve N of the equilibrium D intersects the bifurcation curves $\mathsf{A}_1 C$ and A_1 (Fig. 5.3.15). The values of the parameters in (5.3.15) have been chosen in such a way that the bifurcation curves in the (α_1, α_2)-plane are located qualitatively as shown in Figs. 5.3.12*d* and 5.3.14. The coordinates of the codimension-two bifurcation points in (5.3.15) are:

$$\mathbf{A}_1\mathbf{B}_2\mathbf{C} : \alpha_1 = 4,\ \alpha_2 = 4\,;$$

$$\mathbf{A}_2\mathbf{B}_2\mathbf{C} : \alpha_1 = 1.5, \alpha_2 = 0.25\,.$$

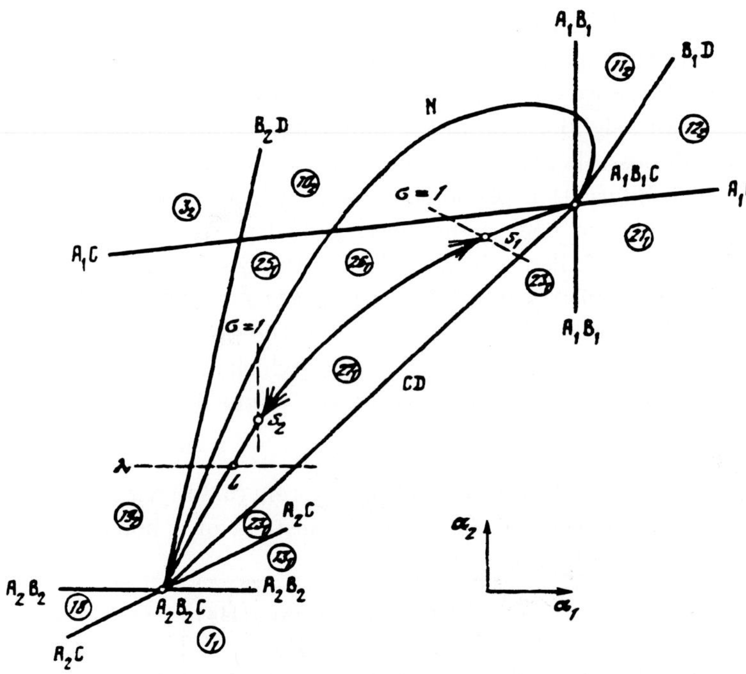

Fig. 5.3.15. Schematic representation of a neighborhood of the heteroclinic cycle curve in the (α_1, α_2)-plane of system (5.3.15). Regions are numbered as in Figs. 5.3.10–5.3.14.

Let us now consider the form of Shil'nikov's conditions in the (α_1, α_2)-plane (Fig. 5.3.15). The curve λ where the point A_2 is a node-focus is the straight horizontal line $\alpha_2 \approx 0.265$, and it intersects the heteroclinic cycle curve at $l(\alpha_1 \approx 1.509)$.

The curve along which $\sigma = 1$ (see (5.3.14)) intersects the heteroclinic cycle curve at the two points $s_1(\alpha_1 \approx 3.67,\ \alpha_2 \approx 3.65)$ and $s_2(\alpha_1 \approx 1.60,\ \alpha_2 \approx 1.27)$. The heteroclinic cycle curve has therefore three different pieces, each of which gives rise to qualitatively different phase portraits when it is intersected as the parameters change.

We consider the bifurcations that occur in the system as α_2 decreases for different fixed α_1.

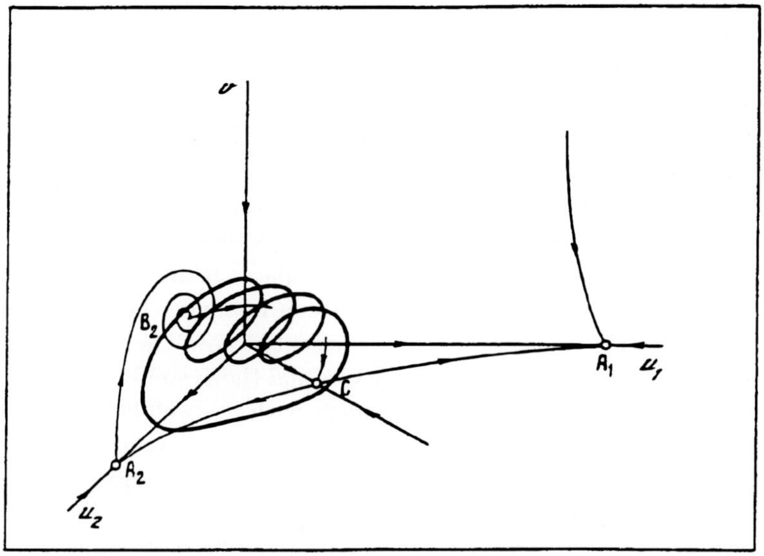

Fig. 5.3.16. Schematic representation of the character of the limit cycle in system (5.3.15) for parameters in a neighborhood of the heteroclinic cycle curve.

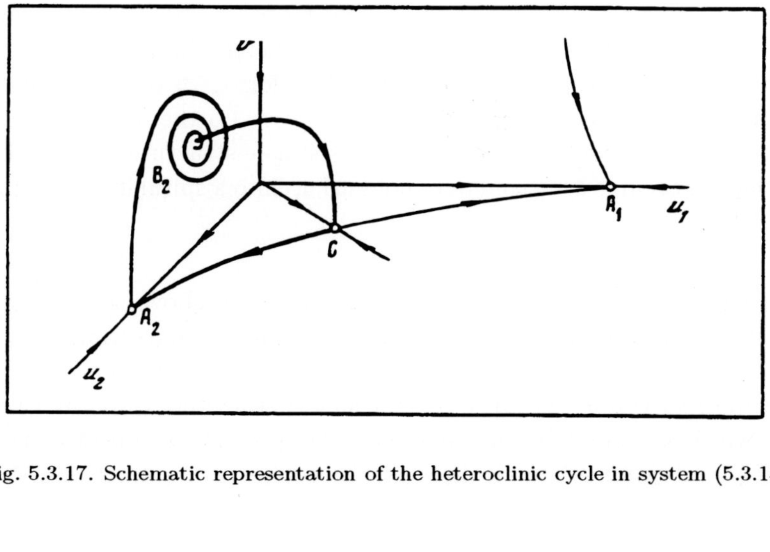

Fig. 5.3.17. Schematic representation of the heteroclinic cycle in system (5.3.15).

1. $4 > \alpha_1 > \alpha_{s_1}$ and $\sigma > 1$, so that the situation corresponds to the first of the above cases. As α_2 decreases a small stable limit cycle appears around the equilibrium D when α_2 intersects the Andronov-Hopf curve of this equilibrium. Further decreasing α_2 lets the limit cycle grow towards the equilibria B_2, A_2 and C, so that its part that passes near the saddle-focus A_2 starts curling up into a spiral (or rather into a helix – Fig. 5.3.16). When α_2 intersects the heteroclinic cycle curve L (Fig. 5.3.17) the limit cycle becomes the heteroclinic cycle and breaks up. (We have not shown the structure of the nontrivial equilibrium D in order to keep the phase portrait simple.) In addition, the equilibrium A_1 becomes globally stable.
2. $\alpha_{s_1} > \alpha_1 > \alpha_{s_2}$. In this interval the bifurcations of the limit cycle that occur when α_2 is decreased may be very complicated. Let us examine a series of bifurcations observed in a numerical experiment for $\alpha_1 = 2.4$ (Fig. 5.3.18). As α_2 decreases we observe a number of successive period doublings of the limit cycle in regions *26* and *27* until, within a certain range of α_2, the numerical experiment gives the impression that the trajectories of the system have completely filled a certain volume of the phase space. It may be said that for those parameter values the system's behavior can no longer be distinguished from random behavior. Using terminology accepted now, the attracting object in phase space is called a strange attractor, and the system behaves chaotically. These numerical results are in agreement with the well-known mathematical findings concerning the existence of this kind of attracting hyperbolic sets near the mentioned values of the parameters. As α_2 decreases further, there is a sequence of decending period doublings of the limit cycle until finally the last stable limit cycle disappears in a saddle-node bifurcation of limit cycles with the limit cycle of saddle-type. Then the equilibrium A_2 becomes globally stable.
3. $\alpha_{s_2} > \alpha_1 > 1.5$. In this case, the bifurcations as α_2 decreases are much the same as in case 1 ($4 > \alpha_1 > \alpha_{s_1}$) with only one difference. The point B_2 is a focus for $\alpha_2 > \alpha_l$ and a node for $\alpha_2 < \alpha_l$. The horizontal line $\alpha_2 = \alpha_l \approx 0.265$ intersects the curve of heteroclinic cycles at the point $l(\alpha_1^l \approx 1.509)$. Accordingly, the bifurcations for $\alpha_{s_2} > \alpha_1 > \alpha_1^l$ are analogous to those for $4 > \alpha_1 > \alpha_{s_1}$, while for $\alpha_1^l > \alpha > 1.5$ the only difference which appears is that the limit cycle retains its simple structure before it breaks up in the heteroclinic cycle and does not curl up into a spiral in the neighborhood of B_2.

Let us consider the ecological interpretation of the above sequences of bifurcations. We have already shown that one of the most interesting behaviors the one-predator–two-preys system can exhibit is that of the predator ensuring the coexistence of two competing prey populations (parameter region and phase portrait *25*), which is impossible without a predator. Let us consider the possible changes of the situation when all three species coexist as α_2 decreases. A decrease

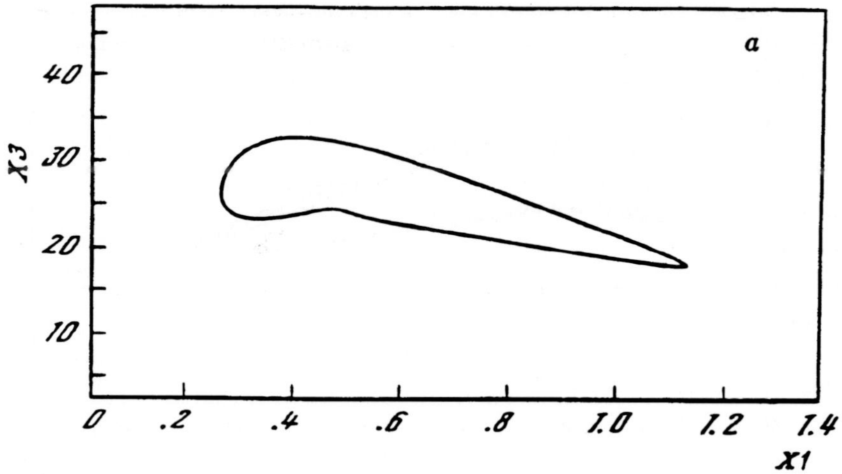

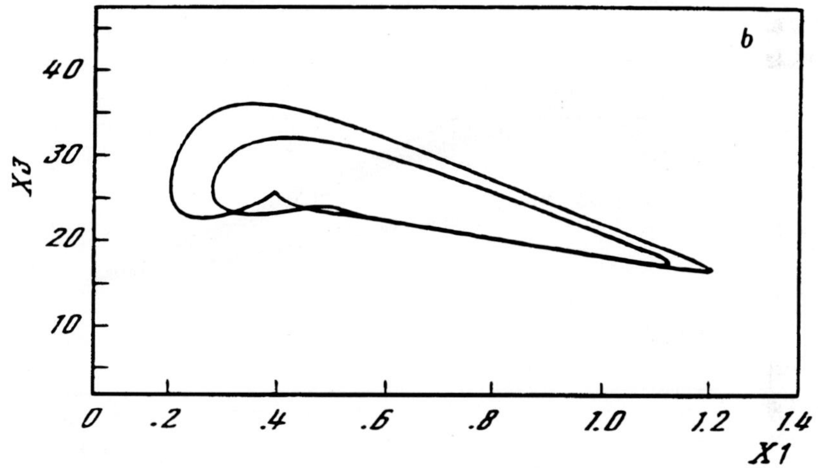

Fig. 5.3.18. Projections of a numerivally calculated three-dimensional attracting trajectory onto the (u_1, v)-plane of system (5.3.15) for fixed $\alpha_1 = 2.4$ and $\alpha_2 = 1.7537$ (*a*), $\alpha_2 = 1.7533$ (*b*), $\alpha_2 = 1.7532$ (*c*), and $\alpha_2 = 1.7531$ (*d*) (Aponina *et al.*, 1982). (Note that the symbols along the axes are $X1 = u_1$ and $X3 = v$).

in α_2 may be interpreted as an increasing load on the community caused by harvesting the second prey species. Whatever the value of α_1, the outcome is always the same: an increase in the load over a certain threshold drives the harvested prey population to extinction, and at a low biotic potential of the first prey population ($\alpha_1 < 4$), the predator dies out, too. In the latter case, the sizes of prey and predator populations do not decrease gradually with the increasing load, but they die out all at once as a result of crossing the dangerous parameter boundary. However, the

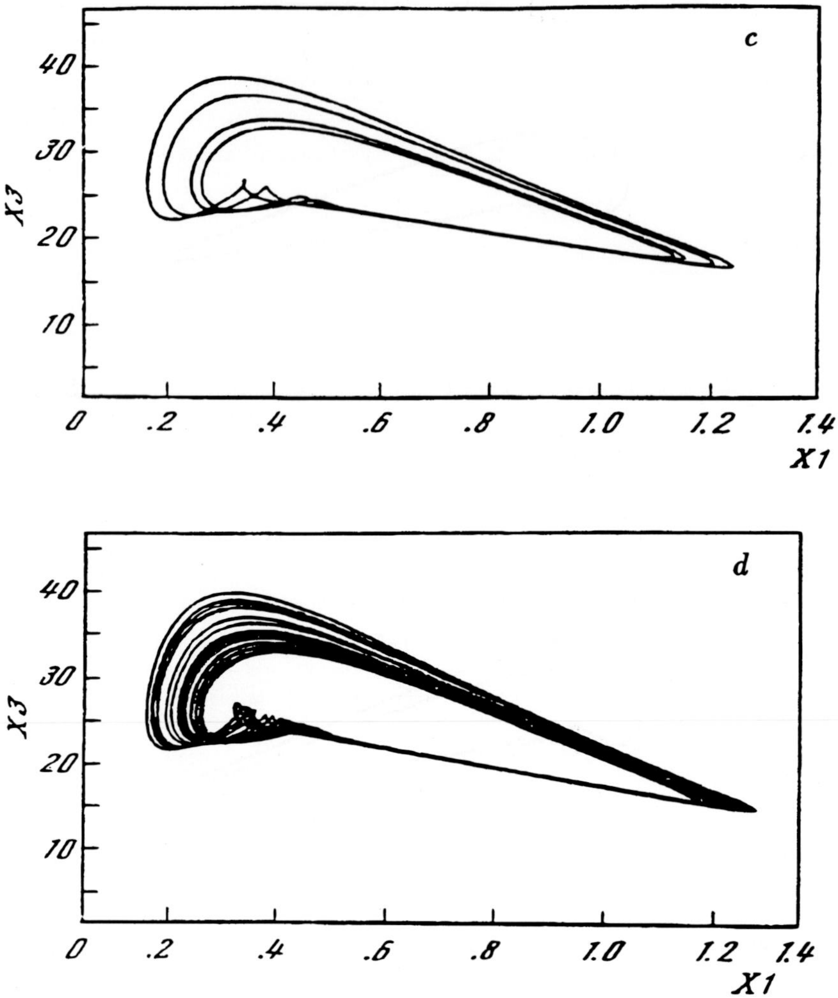

Fig. 5.3.18. (Continued)

character of this boundary and, accordingly, the criteria for approaching it may be different. For the system considered here, the first hint that the dangerous boundary is approaching consists of a gradual excitation of oscillations. Better criteria for approaching the dangerous parameter boundary depend on the value of the biotic potential α_1 of the first prey species. For large and small biotic potentials, such a criterion is the development of a typical relaxation character of the oscillations and, possibly, the complication of the shapes of limit cycles. At intermediate values of the biotic potential α_1, the indication for approaching the dangerous boundary is a series of successive period-doubling bifurcations of the oscillations developing into chaotic behavior.

5.4. Lower Critical Density of the Producer in a System of Three Trophic Levels

What new dynamic effects appear if we introduce the factor of lower threshold density of the producer population in the model of a producer–consumer–predator community? The addition of this factor yields the system (Bazykin, Aponina, 1981):

$$\begin{aligned} \dot{x} &= ax(x-L)(K-x) - b_1xy\,, \\ \dot{y} &= -c_1y + d_1xy - b_2yz\,, \\ \dot{z} &= -c_2z + d_2yz \end{aligned} \tag{5.4.1}$$

Setting $x = Ku$, $y = (aK/b_1)v$, $z = (d_1K/b_2)w$, and $t = \tau/aK$ changes (5.4.1) to

$$\begin{aligned} \dot{u} &= u(u-l)(1-u) - uv\,, \\ \dot{v} &= -\gamma_1 v(m_1 - u + w)\,, \\ \dot{w} &= -\gamma_2 w(m_2 - v) \end{aligned} \tag{5.4.2}$$

where $\gamma_1 = d_1/a$, $\gamma_2 = d_2/b_1$, $m_1 = c_1/d_1K$, $m_2 = c_2b_1/ad_2K$, and $l = L/K$. The equilibria can be found from the corresponding algebraic equations as:

$$\begin{aligned} &\mathbf{O}\ (u = v = w = 0)\,; \\ &\mathbf{A}_1\ (u = l,\ v = w = 0)\,; \\ &\mathbf{A}_2\ (u = 1,\ v = w = 0)\,; \\ &\mathbf{B}\ (u = m_1,\ v = (m_1 - l)(1 - m_1),\ w = 0)\,; \\ &\mathbf{D,E}\ (u_{\mathrm{D,E}} = ({}^1/_2(1 + l \pm \sqrt{(1-l)^2 - 4m_2}\,)\,, \\ &\qquad v_{\mathrm{D,E}} = m_2\,, \\ &\qquad w_{\mathrm{D,E}} = {}^1/_2(1 + l - 2m_1 \pm \sqrt{(1-l)^2 - 4m_2}\,),))\,. \end{aligned} \tag{5.4.3}$$

where the "+" indicates the coordinates of the point D, and the "−" those of the point E.

We start constructing the bifurcation diagram by studying the stability and the bifurcations of equilibria in the coordinate planes of the phase space. The phase portraits in the coordinate planes $\{u = 0\}$ and $\{v = 0\}$ are invariant for all values of the parameters (Fig. 5.4.1).

In the plane $\{u = 0\}$ the origin is a global attractor. In the plane $\{v = 0\}$ the line $\{u = l\}$ is the stable manifold separating the basins of attraction of the origin from that of the point $(u = 1, w = 0)$. The behavior of the system in the plane $\{w = 0\}$ has been studied in detail in Sec. 3.5.5. The bifurcation diagram in the (l, m_1)-plane and a complete set of phase portraits are shown in Figs. 3.5.8 and 3.5.9.

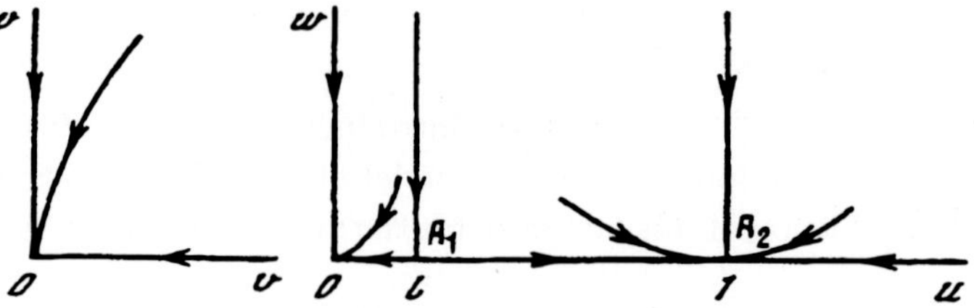

Fig. 5.4.1. Phase portraits of system (5.4.2) in the coordinate planes $\{u = 0\}$ and $\{v = 0\}$.

Let us now turn to the study of the nontrivial equilibria D and E. To that end we determine the boundaries of the parameter regions for which there are two, one ore no nontrivial equilibria at all in the first octant of the phase space. These boundaries are given by the parameters for which the equilibria D and E undergo a saddle-node bifurcation with with the equilibrium B, or with each other:

$$\mathbf{DE}: 4m_2 = (1-l)^2\,,$$
$$\mathbf{BD}: m_2 = (m_1 - l)(1 - m_1), \quad m_1 > {}^1/_2(1+l)\,, \tag{5.4.4}$$
$$\mathbf{BE}: m_2 = (m_1 - l)(1 - m_1), \quad m_2 < {}^1/_2(1+l)\,.$$

If we study the stability of the points D and E we can see that in the first octant D is stable for all values of the parameters ($\lambda_1^{\mathbf{D}} < 0$, Re $\lambda_{2,3}^{\mathbf{D}} < 0$), whereas the point E is always unstable ($\lambda_1^{\mathbf{E}} > 0$, Re $\lambda_{2,3}^{\mathbf{E}} > 0$). In particular, this implies that a bifurcation of conditional codimension one and of true codimension two, due to the concrete form of the system, occurs for $4m_2 = (1-l)^2$ when both, the real eigenvalue and the real part of a pair of complex eigenvalues, are equal to zero simultaneously. Consequently, when crossing this bifurcation curve in parameter space, a small saddle-type limit cycle, as well as the pair of points D and E, appear at the same time.

To get an exact idea of the relative positions of the bifurcation surfaces (5.4.4) in the three-dimensional (m_1, m_2, l)-space it is sufficient to analyse its intersections with the two-dimensional (m_1, m_2)-planes for all $1 > l > 0$. The bifurcations that occur in the plane $\{w = 0\}$ are represented by the vertical straight lines in the (m_1, m_2)-plane (Fig. 5.4.2). A section of the bifurcation surface $\{4m_2 = (1 - l^2)\}$ is shown in the bifurcation diagram only on the left of the line $\{m_1 = (1+l)/2\}$, corresponding to a saddle-node bifurcation of the equilibria D and E in the positive octant of the phase space.

The bifurcations in (5.4.4) do not describe the bifurcation diagram completely. The system has one more (nonlocal) bifurcation, namely the disappearance of the saddle-type limit cycle C_E by "moving through" the stable limit cycle C_B in the plane $\{w = 0\}$. In the bifurcation diagram (Fig. 5.4.2) this bifurcation is the boundary of parameter regions *9* and *12*. As a result, the limit cycle C_E leaves the positive octant, and the limit cycle C_B becomes saddle-type, that is, it remains stable in the plane $\{w = 0\}$, but becomes repelling in the w direction. Since this bifurcation is nonlocal, the corresponding bifurcation curve has no analytic expression. From general considerations it is clear that this bifurcation curve connects the points $(m_1 = l,\ m_2 = 0)$ and $(m_1 = m_{sp},\ m_2 = {}^1/_4(1-l)^2)$. For $l = 0$ and $l = 1$ this curve was numerically traced by E.A. Aponina (Bazykin, Aponina, 1981) with the aid of the computer program CYCLE (Khibnik, 1979). The complete set of phase portraits of the system is also shown in Fig. 5.4.2.

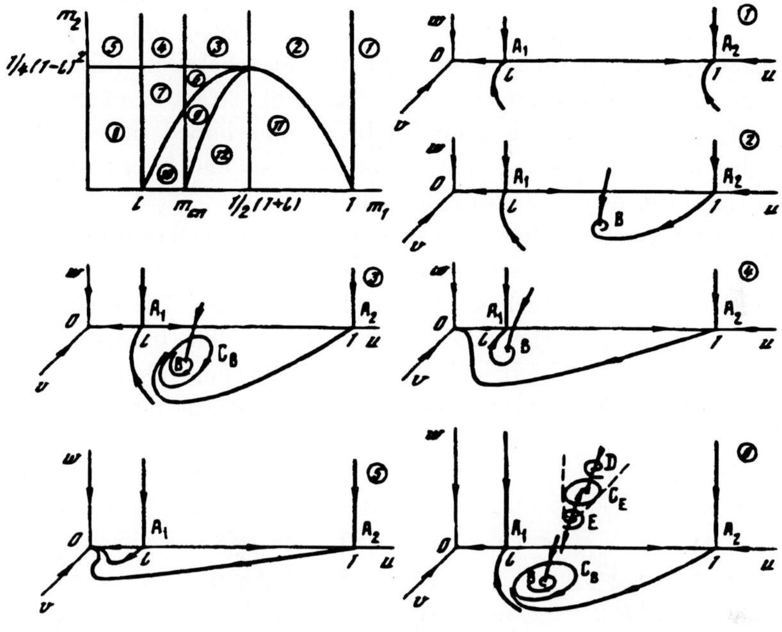

Fig. 5.4.2. Bifurcation diagram and the complete set of twelve phase portraits of system (5.4.2). For parameters from the curve m_{sp} there is a heteroclinic connection in the plane $\{w = 0\}$ between the saddles A_2 and A_1. The boundary of the basin of attraction of D is shown schematically by a dotted curve.

Now we proceed to the ecological interpretation of bifurcation diagram and phase portraits. For parameters from regions *1–5* the coordinate plane $\{w = 0\}$ is an attractor. In other words, in this case the predator always dies out, and the behavior of system (5.4.2) does not differ from that of system (5.3.12). The predator is so poorly adapted to the consumer population ($m_2 < (1-l)^2/4$) that the latter is unable to feed the predator, so that it is always doomed to die out.

In the absence of a predator, the producer and consumer populations can coexist in a stationary (region *11*) or an oscillatory (*12*) regime. The situation is unstable against the invasion of the predator. If it appears in the community, its population grows, and the community reaches the stationary state D of stable coexistence of all three populations. For parameters from region *12* the predator stabilizes the oscillations which have developed in the producer–consumer community in its absence. It should also be noted that for parameters from region *11* the equilibrium density of the consumer in the absence of a predator is higher than when all three populations coexist ($v_\mathbf{B} > v_\mathbf{D}$), while the respective equilibrium density of the producer is lower. This is in full agreement with an intuitive explanation.

For parameters from regions *11* and *12*, as for any value of the parameters, the origin of the phase space is stable. The basins of attraction of the equilibria O and D are separated by the two-dimensional stable manifold of A_1.

For parameters from regions *7*, *8* and *10*, as well as from regions *11* and *12*, two attractors – the origin (extinction of all three populations) and the point D (their stable coexistence) exist in phase space. The difference between these regions lies in the character of the surface separating the basins of attraction of these equilibria. The basin of attraction of D is bounded by a cone-shaped surface which either completely lies in the first octant (parameter regions *7* and *8*) or touches the plane $\{w = 0\}$ at the point B (region *10*) (Fig. 5.4.2).

The greatest diversity of behavior is found for parameters from regions *6* and *9*. For these regions three attractors exist simultaneously – the origin, the equilibrium D and the stable limit cycle C_B in the plane $\{w = 0\}$. The basin of attraction of the origin is bounded by the stable manifold of A_1. The basins of attraction of D and of the stable cycle C_B are separated by a cone-shaped surface. It lies either completely inside the first octant (region *6*) or touches the plane $\{w = 0\}$ at the point B (region *9*).

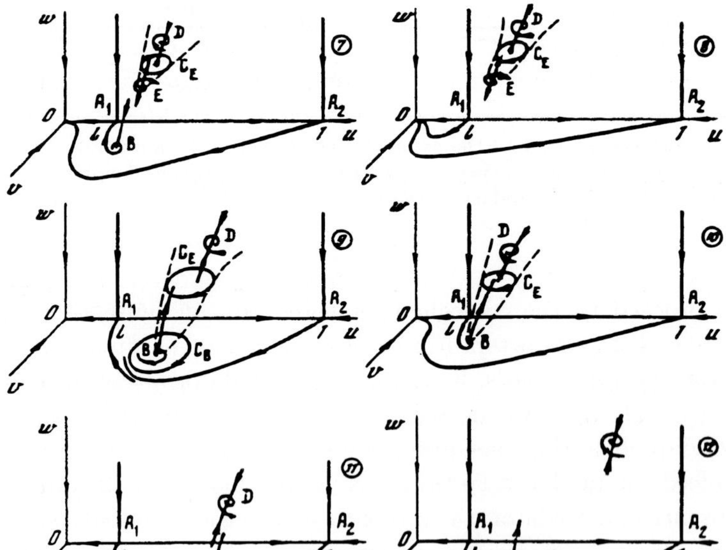

Fig. 5.4.2. (Continued)

We have now completely described all possible phase portraits for different values of the parameter. Now we list the attracting regimes that can occur separately or simultaneously in the system for fixed parameters.

1. The trivial equilibrium O (the origin of the phase space) is stable for all values of the parameter. It corresponds to the extinction of all three populations.
2. The equilibrium A_2 is stable for parameters from region *1*. It corresponds to the isolated existence of the producer population in the absence of the consumer and the predator.
3. The equilibrium B is stable for parameters from region *2*. It corresponds to the stationary coexistence of the producer and the consumer in the absence of the predator.
4. The limit cycle $\mathsf{C}_\mathbf{B}$ is stable for parameters from regions *6* and *9*. It corresponds to the oscillatory coexistence of the producer and the consumer populations in the absence of the predator.
5. The equilibrium D is stable for parameters from regions *6–12*. It corresponds to the stationary coexistence of all three populations.

Let us now consider more closely the basin of attraction of D for parameters from different regions, as well as perturbations that drive the system out of that basin.

It has been shown already that the character of that basin is qualitatively different for parameters from regions *11* and *12* on the one hand, and from regions *6* and *10* on the other hand. In the first case, for $u > l$ and any arbitrarily small initial v and w, the system always returns to D. In the second case, any sufficiently strong perturbation decreasing v and/or w when the system is in D takes the system beyond the boundary of the basin of attraction of D. Then the predator always dies out, and the surviving producer–consumer populations may either coexist in an oscillatory regime (regions *6* and *9*) or also die out.

Now we describe the bifurcations affecting the regime of stationary coexistence of all three populations.

1. For $m_1 > (l+1)/2$ and as the parameter m_2 decreases and passes through $m_2 = (m_1 - l)(1 - m_1)$, the system passes from region *2* into region *11*. The stationary coexistence of the producer and consumer populations becomes unstable against the invasion of the predator, and the equilibrium D appears, corresponding to the stationary coexistence of all three populations. The opposite change of m_2 gradually decreases the density of the predator until it becomes zero at $m_2 = (m_1 - l)(1 - m_1)$, that is, the predator dies out "gradually".
2. For $(1+l)/2 > m_1 > m_{sp}$ and large enough m_2 corresponding to parameter region *3* the only attracting regime is the limit cycle C_B in the plane $\{w = 0\}$, corresponding to the oscillatory coexistence of the producer and consumer populations in the absence of the predator. As m_2 decreases and passes

through the boundary between regions *9* and *12* the cycle C_B suddenly loses its stability, and the system jumps to D. In biological terms this means the following. If a small number of predators invades the community then for the values of m_2 above a certain threshold their population density is maintained at a very low level exclusively due to immigration. When m_2 drops below threshold the predator population density increases suddenly to the level w_D. At the same time, the oscillations of the producer and consumer populations are damped.
If the parameter m_2 moves in the opposite direction the equilibrium D is perturbed in a sudden manner at the same value of m_1 (it disappears in a saddle-node bifurcation with E), and the system jumps to the limit cycle C_B. When the parameter m_2 crosses the threshold the predator dies out, and the remaining producer–consumer community jumps to large-amplitude oscillations. It is important to emphasize that the latter phenomenon occurs for another, larger value of m_2, namely for $m_2 = (1-l)^2/4$ (the boundary between regions *6* and *3*). Consequently, there is a kind of hysteresis.

3. For $m_1 < m_{sp}$ and $m_2 < (1-l)^2/4$ the system shows a stable stationary coexistence of all three populations in the equilibrium D. As m_2 increases the basin of attraction of D shrinks, and for $m_2 = (1-l)^2/4$ the equilibrium is suddenly perturbed. is the same?); the author this change. Then the origin remains as the only attractor and all three populations die out.

Let us sum up the main results of this section. What new events can occur in the producer–consumer community with a threshold density of the producer (see Sec. 3.5.5) if we introduce a third trophic level, namely the predator? For low adaptability of the consumer the predator brings no new events: since the producer is unable to feed the consumer, naturally the predator is doomed to die out. At an average adaptability of the consumer, the predator survives if it is sufficiently adapted to the consumer. In this case, a gradual deterioration of its life conditions causes its equilibrium size to gradually decrease to zero. At a high, but not excessive, adaptability of the consumer, also the predator can survive in the community, and this stabilizes the producer and consumer population densities, which oscillate without the predator.

Of the greatest interest is the case of excessive adaptability of the consumer to the producer. Here, the presence of a predator creates a possibility of stable stationary coexistence of the producer, consumer and predator populations, while without the predator producer and consumer cannot coexist. It should be stressed that in this case both, a considerable single decrease in the predator population size, and the gradual deterioration of the predator's life conditions, leads to the complete extinction of the community. Here, unlike in the case of an intermediate consumer adaptability, a gradual deterioration of the predator's life conditions does not result in a gradual decrease of the predator's equilibrium size to zero. Instead it only changes gradually until a threshold is reached, passing which drives the community to complete extinction.

The case of excessive consumer adaptability to the producer in the three-level community can readily be illustrated by the deliberately schematized example grass–deer–wolf. The extermination of wolves enables the deer to reproduce so much that they eat almost all the grass, after which they die out. Then grass is replaced by shrubs, and an entirely new ecosystem appears.

Even more interesting is the fact that a single abrupt decrease in the consumer population size may lead to similar catastrophic consequences. In this case, a convenient example is the forest–pest-insect–bird community. Here the following events can occur: a considerable single artificial reduction of the size of the destructive insects (the consumer) results in the decrease of the population density of the insectivorous birds (the predator). The weakened predator pressure leads to an outbreak of insects having escaped predation. The insects eat all their resources (needles or foliage) to a level below their threshold. After that the initial forest perishes, and for a long time it will be replaced by a qualitatively different ecosystem.

In fact, the dynamics of natural ecosystems are incomparably more complicated. Nevertheless, despite being utmost and intentionally sketchy, our examples apparently correspond to similar events in reality.

Appendix

A 5.2.1
Phase portraits
of Eq. 5.2.2

$$\frac{dx}{dt} = x(1-z)$$
$$\frac{dy}{dt} = \gamma_1 y(n-z)$$
$$\frac{dz}{dt} = -\gamma_2 z(1-x-y)$$

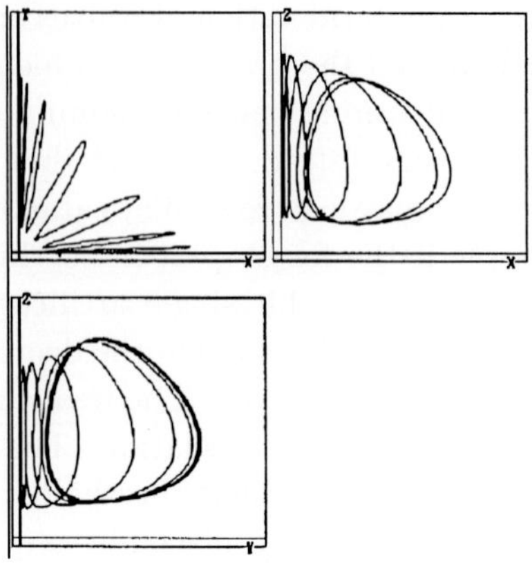

$\gamma_1=1$
$\gamma_2=1$
$n=1.2$

A 5.2.2
Phase portraits
of Eq. 5.2.6

$$\frac{dx}{dt} = x(1-y)$$
$$\frac{dy}{dt} = -\gamma_1 y(1-x+z)$$
$$\frac{dz}{dt} = -\gamma_2 z(\alpha-y)$$

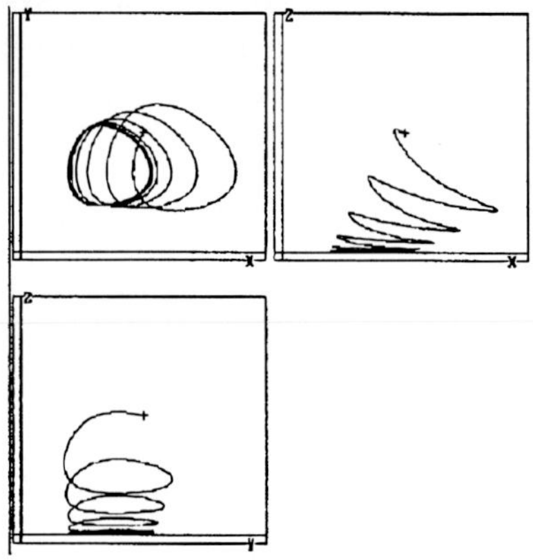

$\gamma_1=1$
$\gamma_2=1$
$\alpha=1.2$

A 5.2.3
Phase portraits
of Eq. 5.2.8

$$\frac{dx}{dt} = x(1-y-\beta z)$$
$$\frac{dy}{dt} = -\gamma_1 y(1-x+z)$$
$$\frac{dz}{dt} = -\gamma_2 z(\alpha-y-\delta x)$$

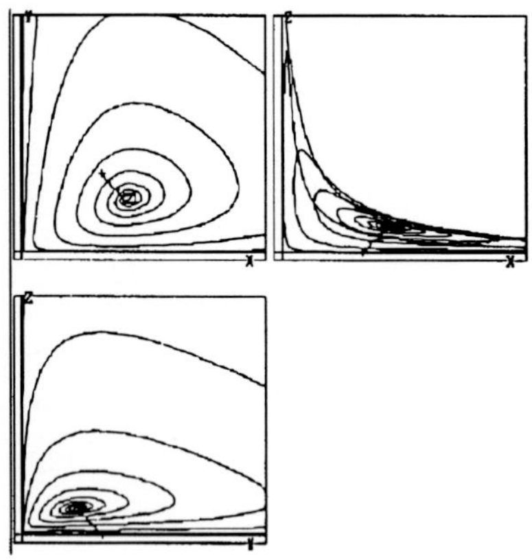

Region 1

$\gamma_1=1$
$\gamma_2=1$
$\alpha=1$
$\beta=1$
$\delta=0.25$

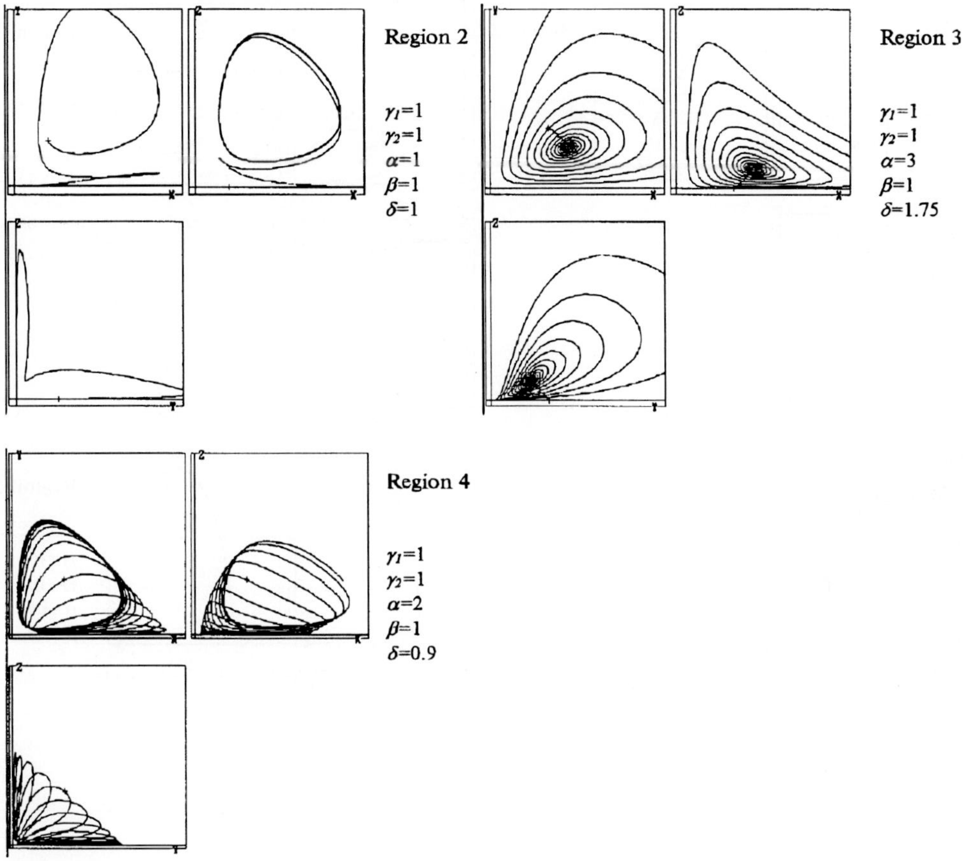

A 5.3.1
Phase portraits of Eq. 5.3.2

$$\frac{dx}{dt} = x(1 - y - \varepsilon x)$$

$$\frac{dy}{dt} = -\gamma_1 y(1 - x + z)$$

$$\frac{dz}{dt} = -\gamma_2 z(\alpha - y)$$

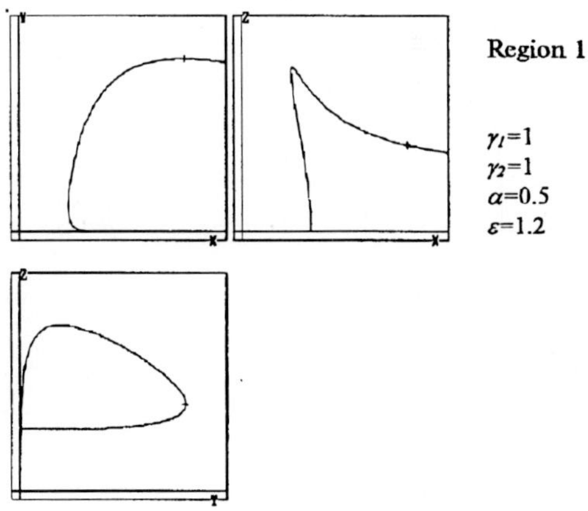

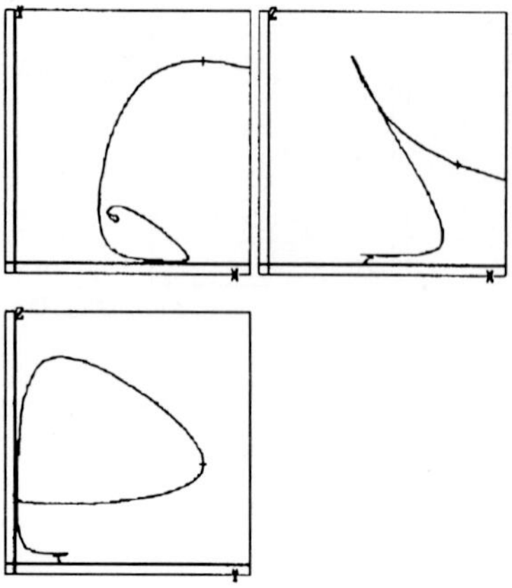

Region 2

$\gamma_1=1$
$\gamma_2=1$
$\alpha=0.5$
$\varepsilon=0.51$

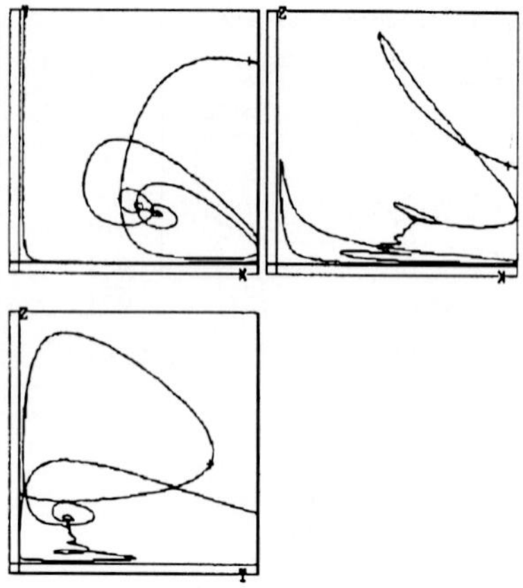

Region 3

$\gamma_1=1$
$\gamma_2=1$
$\alpha=0.5$
$\varepsilon=0.32$

A 5.3.2
Phase portraits
of Eq. 5.3.4

$$\frac{dx}{dt}=x(1-y-z-x)$$
$$\frac{dy}{dt}=-\gamma_1 y(\alpha-x)$$
$$\frac{dz}{dt}=-\gamma_2 z(\beta-x)$$

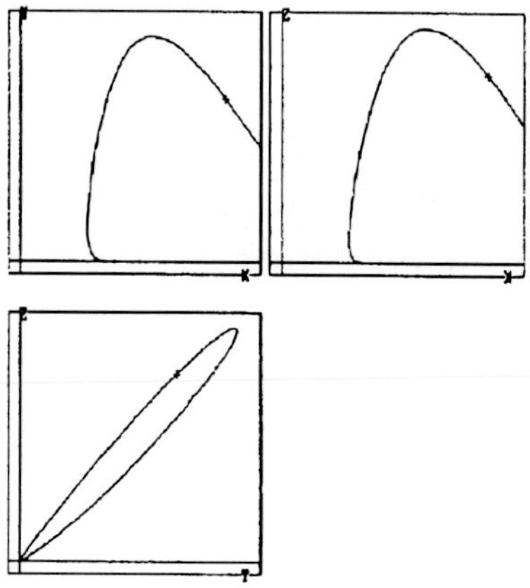

Region 1

$\gamma_1=10$
$\gamma_2=10$
$\alpha=1.1$
$\beta=1.2$

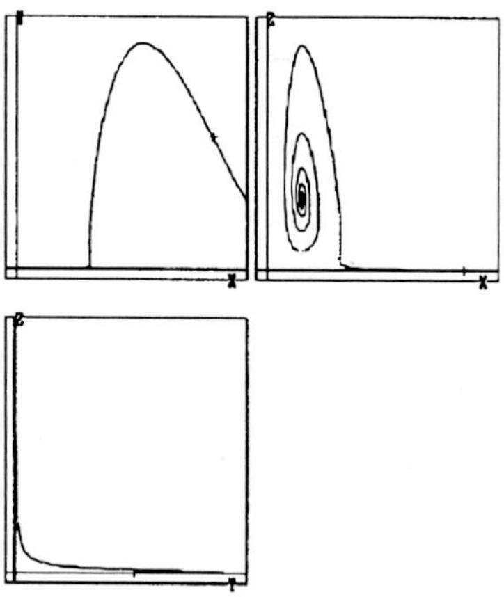

Region 3

$\gamma_1=10$
$\gamma_2=10$
$\alpha=1.1$
$\beta=0.3$

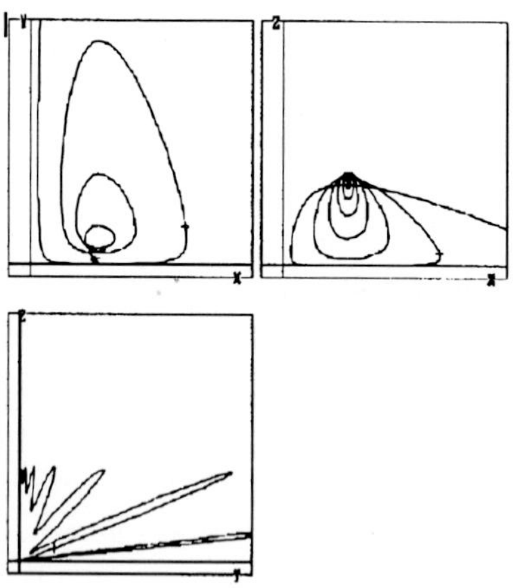

Region 5

$\gamma_1=10$
$\gamma_2=10$
$\alpha=0.31$
$\beta=0.29$

A 5.3.3
Phase portraits
of Eq. 5.3.6

$$\frac{dx}{dt} = x(1 - y - \beta z - \varepsilon x)$$

$$\frac{dy}{dt} = -\gamma_1 y(1 - x + z)$$

$$\frac{dz}{dt} = -\gamma_2 z(\alpha - y - \delta x)$$

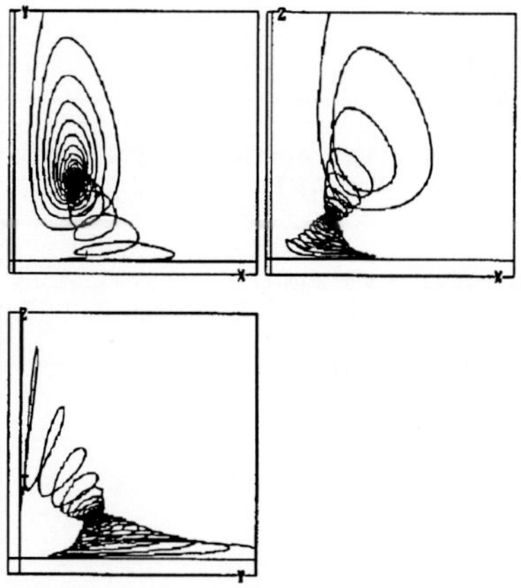

Region 1

$\gamma_1=1$
$\gamma_2=1$
$\alpha=1.5$
$\beta=1$
$\delta=0.7$
$\varepsilon=0.1$

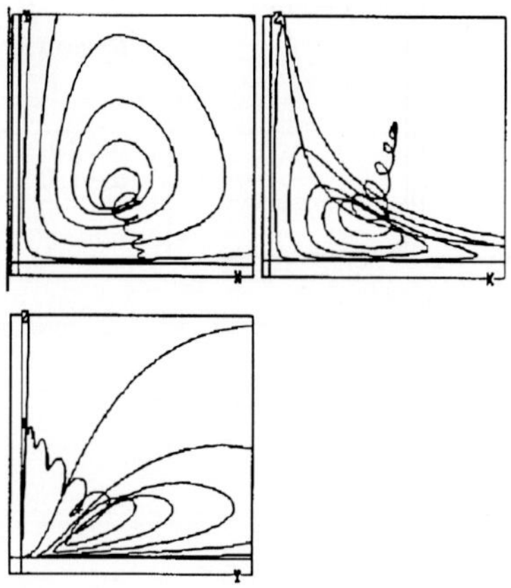

Region 3

$\gamma_1=1$
$\gamma_2=1$
$\alpha=2.4$
$\beta=1$
$\delta=1.5$
$\varepsilon=0.1$

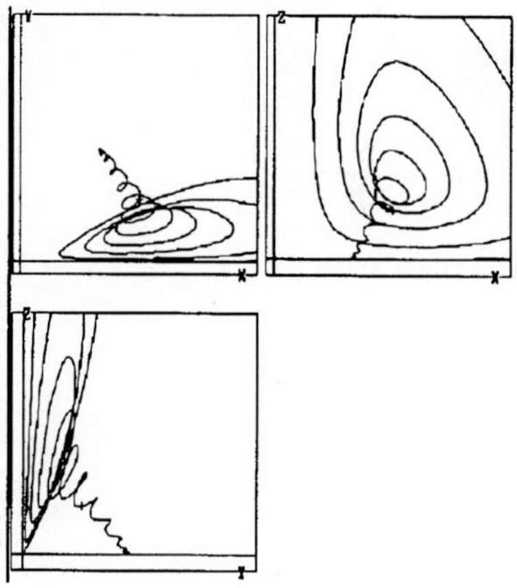

Region 4

$\gamma_1=1$
$\gamma_2=1$
$\alpha=2.4$
$\beta=1$
$\delta=1.3$
$\varepsilon=0.1$

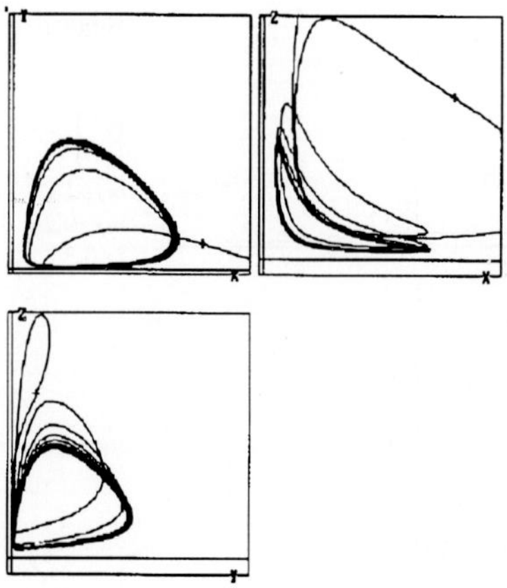

Region 7

$\gamma_1=1$
$\gamma_2=1$
$\alpha=1$
$\beta=1$
$\delta=0.3$
$\varepsilon=0.1$

A 5.3.4

Phase portraits
of Eq. 5.3.8

$$\frac{dx}{dt} = x(\alpha_1 - \beta_1 z - x - \varepsilon_1 y)$$

$$\frac{dy}{dt} = y(\alpha_2 - \beta_2 z - y - \varepsilon_2 x)$$

$$\frac{dz}{dt} = -z(1 - \delta_1 x - \delta_2 y)$$

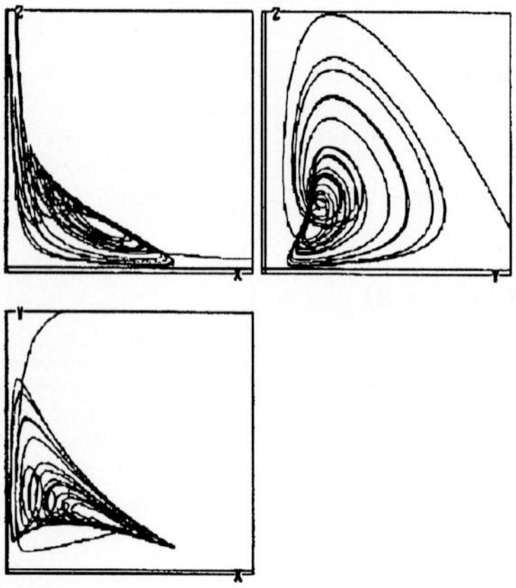

Region 26

α_1=1.363
α_2=1.147
β_1=1
β_2=5
δ_1=0.556
δ_2=5
ε_1=5
ε_2=1

Phase portraits
of Eq. 5.3.15

$$\frac{dx}{dt} = x(\alpha_1 - 4z - x - 6y)$$

$$\frac{dy}{dt} = y(\alpha_2 - 10z - y - x)$$

$$\frac{dz}{dt} = -z(1 - 0.25x - 4y + z)$$

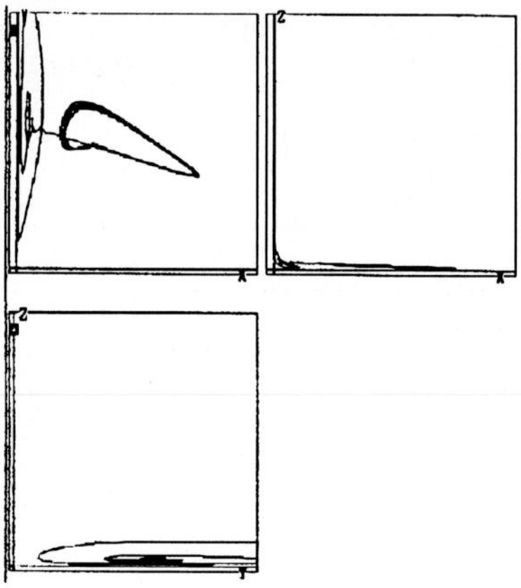

Region 1

α_1=2.4
α_2=1.7537

A 5.4

Phase portraits
of Eq. 5.4.2

$$\frac{dx}{dt} = x(x - l)(1 - x) - xy$$

$$\frac{dy}{dt} = -\gamma_1 y(m_1 - x + z)$$

$$\frac{dz}{dt} = -\gamma_2 z(m_2 - y)$$

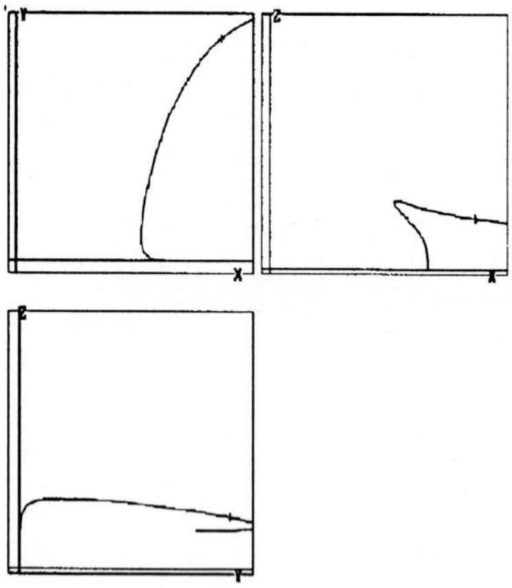

Region 1

γ_1=1
γ_2=1
m_1=1.2
m_2=0.2
l=0.3

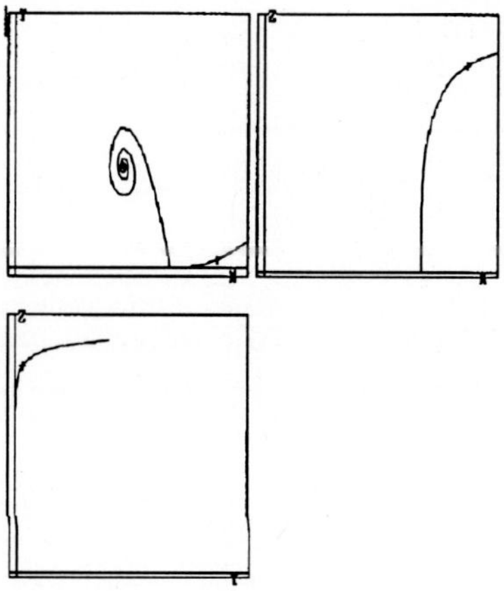

Region 2

$\gamma_1=1$
$\gamma_2=1$
$m_1=0.7$
$m_2=0.2$
$l=0.3$

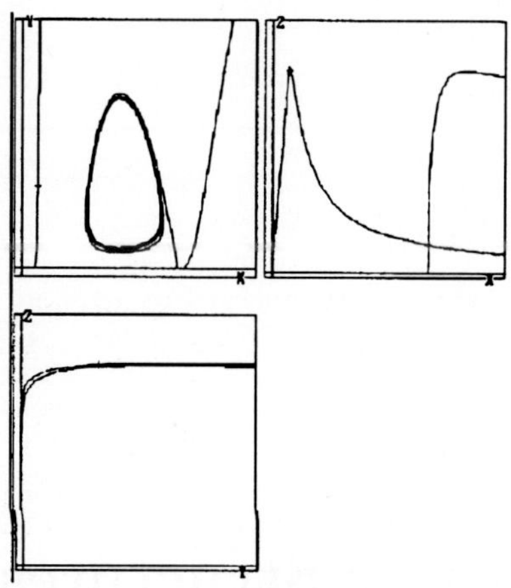

Region 3

$\gamma_1=1$
$\gamma_2=1$
$m_1=0.63$
$m_2=0.2$
$l=0.3$

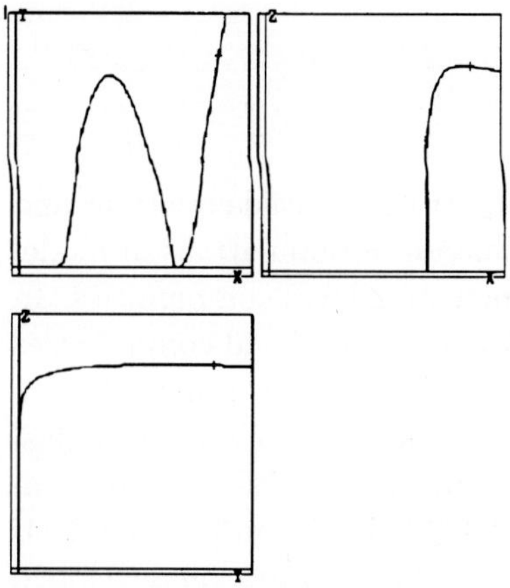

Region 4

$\gamma_1=1$
$\gamma_2=1$
$m_1=0.61$
$m_2=0.2$
$l=0.3$

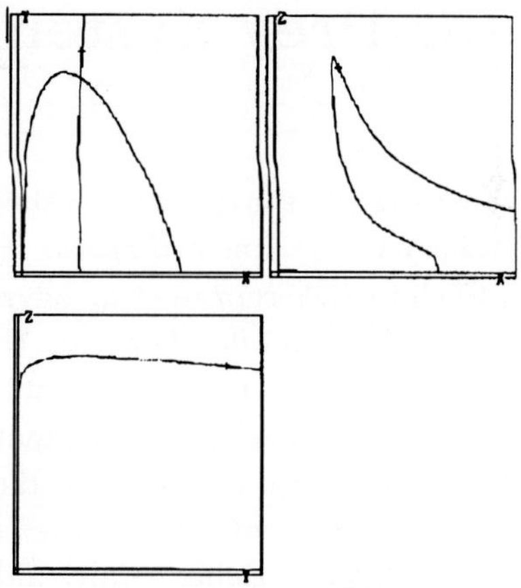

Region 5

$\gamma_1=1$
$\gamma_2=1$
$m_1=0.29$
$m_2=0.2$
$l=0.3$

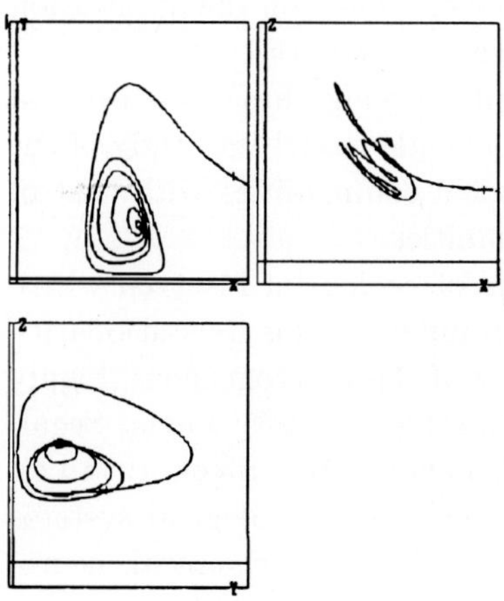

Region 8

$\gamma_1=1$
$\gamma_2=1$
$m_1=0.29$
$m_2=0.1$
$l=0.3$

Chapter 6

Dissipative Structures in Predator-Prey Systems

In this chapter we apply the qualitative theory of ordinary differential equations and bifurcation theory to the study of *spatio-temporal* behavior of populations in model ecosystems. Such models represent an aggregate of migrating local communities. In the simplest case this is a *bilocal* system, that is, a system of two local communities related by migrants between the two populations constituting the communities.

Until recently, the development of mathematical models for solving ecological problems proceeded in two independent directions (Levin, 1980): the study of temporal dynamics of ecological communities (the topic of this book) and the analysis of their spatial structure. Being totally independent, the latter direction developed mainly to satisfy the needs of geobotanics with special emphasis on the application of statistical methods describing the structure of vegetative covers.

In the last decade first steps were made toward unifying these two isolated directions. This means combining in the analysis of a single model the study of the dynamics of spatially homogeneous (or local) ecological communities with that of the stationary spatial structure of distributed communities.

This was suggested by practicioneers from fields of ecological research where neither the abstraction of spatial homogeneity of a community nor its stationarity are adequate. This, first of all, is related to the study of the heterogeneous, highly dynamic structure of plankton distribution in oceans (Dupois, 1975; Levin, Segal, 1976; Masayasu, 1979; Segal, Jackson, 1972; Segal, Levin, 1973; Steel, 1974*a*,*b*). Apart from this, the interest in a spatio-temporal analysis of ecological systems was greatly stimulated by successes of spatio-temporal studies of chemical, neurophysiological and other systems, that is, by research in the new field of synergetics (Haken, 1978).

Mathematical models for studying spatio-temporal dynamics in distributed active media are provided by systems of nonlinear partial differential equations, normally of diffusion type (Turing, 1952). In this book we do not study actual distributed systems described by diffusion equations with nonlinear kinetic terms. We restrict ourselves to the study of a bilocal system because, first, such a study is of independent ecological interest and, second, the results obtained may help understand the dynamics of real distributed systems. (A similar approach was used, for instance, by Hilborn (1979), Sullivan (1988), Antonovsky *et al.* (1990).)

Of the wide diversity of phenomena that may occur in spatially distributed ecological systems, we dwell only on stationary dissipative structures and their relationship with the temporal dynamics of the corresponding local systems. As a consequence, we exclude from our consideration a range of interesting problems related to the propagation of nonlinear waves ("life waves" could be one possible meaning of this common term) in distributed ecological systems.

The last section of this chapter is devoted to the study of a model of a stationary dissipative structure appearing in a bilocal ecological system as a result of Darwinian evolution of one of the species of the community.

6.1. Bilocal System

Here we study the possible dynamics in a model ecological community that consists of two local predator–prey systems connected by a two-way influx of migrants. Our main interest here is the relationship between the oscillations in appropriate local predator–prey systems and the stationary spatially heterogeneous regimes in the bilocal system. Therefore, we consider a system of differential equations of the form

$$\begin{aligned} \dot{u} &= f(u, v, \alpha)\,, \\ \dot{v} &= g(u, v, \alpha)\,, \end{aligned} \tag{6.1.1}$$

and such that in a certain range of the parameters α the system has a stable limit cycle.

We are mainly interested in a bilocal system as a simple method of describing a spatially distributed community. The initial system given in the local case by (6.1.1), when written with partial derivatives describing spatio-temporal dynamics, becomes

$$\begin{aligned} \frac{\partial u}{\partial t} &= D_u \frac{\partial^2 u}{\partial r^2} + f(u, v, \alpha)\,, \\ \frac{\partial v}{\partial t} &= D_v \frac{\partial^2 u}{\partial r^2} + g(u, v, \alpha)\,, \end{aligned} \tag{6.1.2}$$

where D_u and D_v are the diffusion coefficients for prey and predator, respectively. The variable r varies on the interval $[0,\ 1]$, where we impose no-influx (or Neuman) boundary conditions:

$$\left.\frac{\partial v}{\partial r}\right|_{r=0, r=1} = \left.\frac{\partial u}{\partial r}\right|_{r=0, r=1} = 0\,.$$

The corresponding bilocal system takes the form (Bazykin, Khibnik *et al.*, 1980)

$$\begin{aligned}\dot u_1 &= f(u_1, v_1, \alpha) - m_u(u_1 - u_2)\,,\\ \dot u_2 &= f(u_2, v_2, \alpha) - m_u(u_2 - u_1)\,,\\ \dot v_1 &= g(u_1, v_1, \alpha) - m_v(v_1 - v_2)\,,\\ \dot v_2 &= g(u_2, v_2, \alpha) - m_v(v_2 - v_1)\,,\end{aligned} \tag{6.1.3}$$

where the indices 1 and 2 refer to the densities of the populations in the first and in the second local subsystems, respectively. The coefficients m_u and m_v describe the migration rates between prey and predator, respectively. They are the analogues of the diffusion coefficients in system (6.1.2).

Experience suggests that dissipative structures in spatially distributed systems appear usually when various components of the system have greatly differing diffusion coefficients. Therefore, we focus our attention on the study of the limit case of a large coefficient of migration for one of the components of (6.1.3). For ecological reasons we consider the case $m_v \gg 1$ when the predator is very mobile, so that total mixing occurs rapidly ($v_1 \approx v_2$). Then system (6.1.3) can be reduced to the three-dimensional system

$$\begin{aligned}\dot u_1 &= f(u_1, v, \alpha) - m(u_1 - u_2)\,,\\ \dot u_2 &= f(u_2, v, \alpha) - m(u_2 - u_1)\,,\\ \dot v &= {}^1/_2\big(g(u_1, v, \alpha) + g(u_2, v, \alpha)\big)\,,\end{aligned} \tag{6.1.4}$$

where $v = (v_1 + v_2)/2$ and $m = m_u$.

Here, any modification of the classical Volterra–Lotka model, in which not more than one nontrivial solution exists for all values of the parameters, can be taken as a concrete system of type (6.1.1). It suffices if in one region of parameters the nontrivial solution is stable, and in the other it is unstable, where its stability is lost gradually in an Andronov-Hopf bifurcation, which gives rise to a small stable limit cycle.

Let us now consider the predator–prey system

$$\begin{aligned}\dot x &= ax^2\,\frac{K - x}{K} - bxy\,,\\ \dot y &= -cy + dxy - hy^2\,,\end{aligned} \tag{6.1.5}$$

which allows for nonlinear reproduction of the prey population at small density, and for intraspecies competition between the prey and the predator. The choice of this form was guided purely by technical reasons, with the intention of simplifying the qualitative study of (6.1.4).

Note that system (6.1.4) in its very nature is invariant under exchanging the indices 1 and 2. In particular, this implies that the plane $\{u_1 = u_2\}$ is a plane of symmetry of the phase space. Therefore, it is natural that the sought for stationary dissipative structures of the distributed system (6.1.2) in the bilocal model (6.1.4) should be compared with equilibria lying outside the invariant plane $\{u_1 = u_2\}$.

Setting $t = \tau/aK$, $x = Ku$, and $y = (aK/b)v$ changes (6.1.5) to

$$\dot{u} = u^2(1-u) - uv\,,$$
$$\dot{v} = -\gamma v(\alpha - u + \delta v)\,, \qquad (6.1.6.)$$

where $\gamma = d/a$, $\alpha = c/dK$, $\delta = ae/bd$.

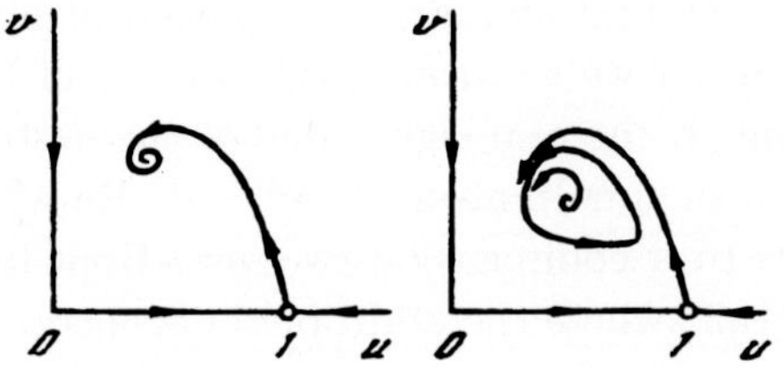

Fig. 6.1.1. Phase portrait of system (6.1.6) for $1 > \alpha > (1-\delta)^2/(1-2\delta)$ (left), and for $(1-\delta)^2/(1-2\delta) > \alpha > 0$ (right). The phase portraits are also those of (6.3.2) for $1/2 < \alpha < 1$ (left) and $\alpha < 1/2$ (right).

Figure 6.1.1 shows phase portraits for different values of the parameters α and δ. The qualitative behavior of system (6.1.6) is independent of γ. From now on we limit ourselves to studying how the system depends on α, for $\gamma = 1$ and $\delta < 1$. For $\alpha > 1$ the behavior is not interesting, since there is no nontrivial equilibrium. Such an equilibrium exists and is stable for $1 > \alpha > (1-\delta)^2/(2-\delta)^2$. When α crosses the threshold $\alpha = (1-\delta)^2/(2-\delta)^2$, the equilibrium loses its stability in an Andronov-Hopf bifurcation, giving rise to a small stable limit cycle.

Thus, we study (6.1.4) in the following form (Bazykin, Khibnik, 1982; Bazykin, Khibnik *et al.*, 1980; Bazykin, Khibnik, Aponina, 1983):

$$\dot{u}_1 = u_1^2(1-u_1) - u_1 v - m(u_1 - u_2)\,,$$
$$\dot{u}_2 = u_2^2(1-u_2) - u_2 v - m(u_2 - u_1)\,, \qquad (6.1.7)$$
$$\dot{v} = v\left(\frac{u_1+u_2}{2} - \alpha - \delta v\right).$$

In the bifurcation diagram of system (Fig. 6.1.2) the positive quadrant of parameter space is divided into nine regions corresponding to qualitatively different behavior. The boundaries of these regions are the curves representing codimension-one bifurcations. The phase portraits corresponding to the regions (*1–8*) are shown in Fig. 6.1.3.

Let us now describe the phase portraits and their bifurcations for different values of the parameters. For $\alpha > 1$ (region *0*) there is only one stable equilibrium

($u_1 = u_2 = 1$; $v = 0$) corresponding to the extinction of the predator. For parameters from region *1* the system has a unique nontrivial globally stable equilibrium A at which predator and prey coexist in a stationary regime. If the migration coefficient m is large enough, a decrease in α drives the system into region *2*. Then the equilibrium A gradually loses its stability, and a stable limit cycle Γ_A appears in the phase portrait. This limit cycle corresponds to an oscillatory regime in which the prey population densities in both subpopulations oscillate synchronously and identically. The phase portraits and bifurcations of the bilocal system (6.1.7) described up to now are identical to those of the local system (6.1.6).

For small values of m the bilocal system shows much greater behavioral diversity than the local one. Let us first examine a sequence of bifurcations of the phase portraits which occurs when we go around the point I in the bifurcation diagram (Fig. 6.1.2). At the point I, the real eigenvalue of the system and the real part of a pair of eigenvalues vanish simultaneously ($\lambda_1^A = 0$, Re $\lambda_{2,3}^A = 0$). In other words, the point I corresponds to a codimension-two zero-Hopf bifurcation of the three-dimensional phase portrait, where the additional degeneracy is due to the system's symmetry.

When the parameters move from region *2* into region *3* the eigenvalue λ_1^A becomes positive and the point A splits into a pair of saddle points B_1 and B_2, for which $\lambda_1^B < 0$ and Re $\lambda_{2,3}^B > 0$. As before (for region *2*), the stable limit cycle Γ_A in the plane $\{u_1 = u_2\}$ remains the only attractor.

When the parameters move into region *4* Re $\lambda_{2,3}^B$ become negative and the equilibria B_1 and B_2 become stable. In this case, the equilibria B_1 and B_2 give rise to limit cycles Γ_1 and Γ_2 of saddle-type, respectively, that is, a subcritical Andronov–Hopf bifurcation occurs. The stable manifolds (which are surfaces) of the limit cycles Γ_1 and Γ_2 separate the basins of attraction of B_1 and B_2 from the basin of attraction of the stable limit cycle Γ_A. As was mentioned above, the stable equilibria B_1 and B_2 lying outside the plane $\{u_1 = u_2\}$ are, within the framework of the stated problem, the analogs of stationary dissipative structures.

When the parameters move from region *4* into region *5* the saddle-type limit cycles Γ_1 and Γ_2 undergo a pitchfork bifurcation and disappear by merging with the stable limit cycle Γ_A. As a result, Γ_A changes to saddle-type. Now only the equilibria B_1 and B_2 (the "dissipative structures") are attractors in phase space. The plane $\{u_1 = u_2\}$ becomes the surface that separates their basins of attraction.

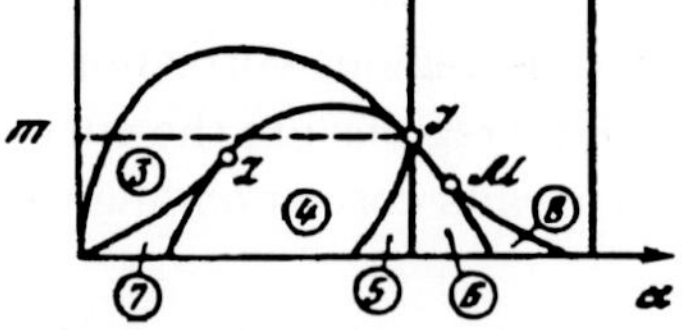

Fig. 6.1.2. Bifurcation diagram of system (6.1.7).

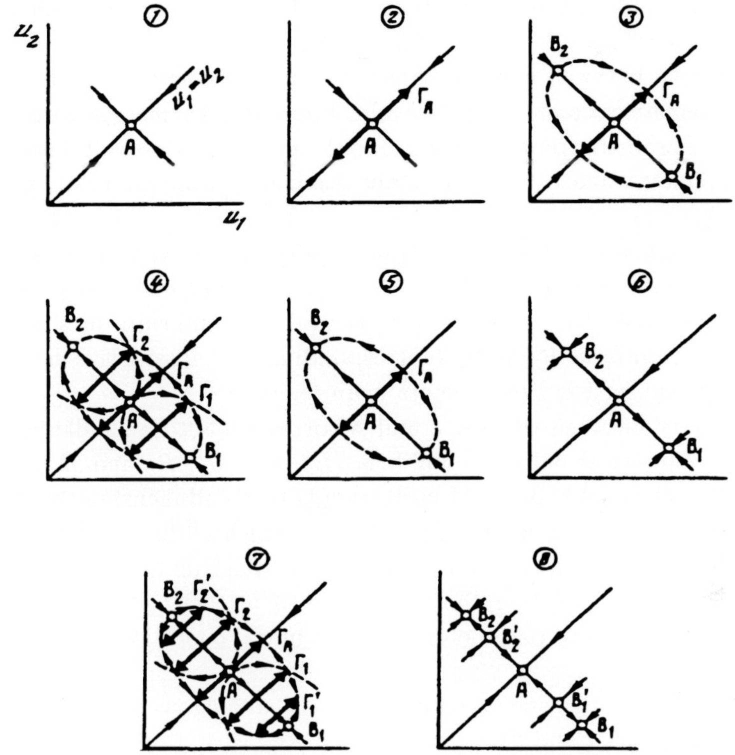

Fig. 6.1.3. Schematic representation of three-dimensional phase portraits of system (6.1.7) for corresponding regions of the bifurcation diagram. The v-axis is perpendicular to the plane of this figure. Solid curves are projections of limit cycles, while dashed curves correspond to invariant surfaces.

When the parameters move from region *5* into region *6* the equilibrium A undergoes an inverse Andronov-Hopf bifurcation: the cycle Γ_A shrinks to A, and the repeller A becomes a saddle with $\lambda_1^A > 0$, Re $\lambda_{2,3}^A < 0$. Finally, when the parameters move back into region *1* the sequence of bifurcations around I is complete: the equilibria B_1 and B_2 disappear in a pitchfork bifurcation with A and, as a result, A becomes globally stable.

Apart from the regions which have the point I in their boundary, the bifurcation diagram shows two more regions, *7* and *8* (Fig. 6.1.3). For parameters from region *7* there simultaneously exist five limit cycles in the phase portrait. Two of them (Γ_1 and Γ_2) are saddle-type limit cycles, and the three limit cycles (Γ_A, Γ'_1 and Γ'_2) are stable. The limit cycle Γ_A corresponds as before to synchronous and identical oscillations of the prey population densities in both subsystems. The stable limit cycles Γ'_1 and Γ'_2 correspond to oscillations of the prey population densities about different equilibria. We call this regime a *nonstationary dissipative structure.*

The boundary between regions *7* and *4* is an Andronov-Hopf bifurcation curve of the equilibria B_1 and B_2: when the parameters move from region *4* into region *7* the equilibria B_1 and B_2 lose their stability gradually, producing small stable limit cycles Γ_1 and Γ_2.

The boundary between regions *7* and *3* is a saddle-node of limit cycle curve, where two symmetrical pairs of limit cycles bifurcate: Γ_1 merges with Γ'_1 and Γ_2 with Γ'_2. The parameter point Z represents the codimension-two bifurcation where the first Lyapunov value L_1 is zero on the Andronov-Hopf curve of the equilibria B_1 and B_2.

For parameters from region *8* the phase portrait shows two saddles (B'_1 and B'_2) and three stable equilibria (A, B_1 and B_2). In other words, depending on the initial condition both, the equilibrium A (a stable equilibrium in a local system) and one of the equilibria B_1 or B_2 (corresponding to a dissipative structure), can exist in the system for fixed parameters from region *8*.

The boundary between regions *1* and *8* corresponds to a saddle-node bifurcation of two symmetrical pairs of equilibria (B'_1 and B_1, B_2 and B'_1) which then disappear. The parameter point M corresponds to a codimension-two bifurcation, namely a degenerate saddle-node bifurcation, in which all five equilibria disappear.

We conclude the description of the bifurcation diagram and the phase portraits of system (6.1.7) by indicating the four kinds of attracting regimes that can be realized, depending on the parameters and the initial conditions. They are

1. The spatially homogeneous equilibrium A – analogous to the one of the local system (regions *1* and *8*).
2. The spatially homogeneous oscillations given by the stable limit cycle Γ_A – analogous to oscillations in the local system (regions *2, 3, 4* and *7*).
3. The stationary dissipative structure given by the points B_1 and B_2 (regions *4, 5, 6* and *8*).
4. The nonstationary dissipative structure given by the stable limit cycles Γ'_1 and Γ'_1 (region *7*). In the terminology suggested by V.A. Vasil'yev (1976), this is a *dynamical dissipative structure.*

Gradual and abrupt rearrangements of the attracting regimes, which occur when the parameter values are changed, can be seen in Figs. 6.1.2 and 6.1.3, and have been partly described above. In particular, system (6.1.7) may exhibit both gradual and abrupt excitation of dissipative structures. The first case corresponds to a direct transition from region *1* into *6*. The second case (hard-excitated dissipative structures) corresponds either to the *1* $\rightarrow$ *6* transition via region *8*, or to passing through the whole sequence of regions *1* $\rightarrow$ *2* $\rightarrow$ *3* $\rightarrow$ *4* $\rightarrow$*5*. (Here oscillations are gradually excited from the equilibrium and then suddenly break up such that the system jumps to a stationary dissipative structure).

Mathematically, this study is based on the investigation of the bifurcations of a system of the form (6.1.4) in a neighborhood of a zero-Hopf point. By this we mean the point of intersection of the bifurcation curves representing one zero and a pair of purely imaginary eigenvalues, respectively (Bazykin, Khibnik, 1979).

In this connection, two questions arise: first, to what extent does the consideration above relate to systems of type (6.1.4) and, second, to what extent is it typical only for the concrete system (6.4.7).

The answer is the following. In a bilocal kinetics-migration system, described by the differential equations (6.1.4), the spatially homogeneous equilibrium may, in the generic case, lose its stability in two ways under variation of the parameters. First, the real part of a pair of complex eigenvalues of the system becomes zero and then positive, which results in an oscillatory loss of stability like in the corresponding local system. The oscillatory loss of stability is necessarily gradual, as above, but in general may be hard as well. Second, the real eigenvalue of the system passes through zero and then becomes positive, giving rise to diffusion loss of stability, and the system may generate a dissipative structure (both gradually and abrupt).

If a system depends on two or more parameters, it is reasonable to describe the combination of oscillatory and diffusion stability losses. In this case, we should study the bifurcations in a neighborhood of the point of intersection of the two curves representing oscillatory and diffusion losses of stability in the parameter plane.

The structure of the parameter neighborhood of this point in the generic case is unknown to the author However, it is clear that apart from the above mentioned curves, at least two more codimension-one bifurcation curves should arrive at this point in the parameter plane. They are the curve of oscillatory loss of stability of the spatially nonhomogeneous equilibria, and the curve of changing stability-type of a homogeneous limit cycle. The relative positions of all bifurcation curves in a neighborhood of the zero-Hopf point give a complete set of phase portraits of the system and their bifurcations. There are several possibilities for different locations of the bifurcation curves, giving a rich picture of unfoldings.

In conclusion, it can be said that in system (6.1.7) we found one example of an unfolding of the zero-Hopf bifurcation, determined by the relative positions of bifurcation curves in a neighborhood of the point I. This makes it possible to distinguish *all* four kinds of attracting regimes that can occur in a bilocal system of the form (6.1.4).

Another question is: How does the analysed behavior of a bilocal system relate to the actual distributed system described by the differential equations of diffusion-kinetics type (6.1.1)? An attempt at sketching a possible answer is made in the following section.

6.2. Annular Habitat

It is a hypothesis that a bilocal model can be used to predict values of the parameters, for which a distributed system exhibits certain behavior. One of the most interesting behaviors of the bilocal system (6.1.7) is the multistable regime (region *4*) in which, depending on the initial condition, the system either generates a stationary dissipative structure, or both its subsystems start oscillating synchronously and identically. Here, we numerically check whether knowing parameter region *4* of the bilocal system (6.1.7) is sufficient for predicting values of the parameters for which there is the analogous regime of multistability in a distributed system (a dissipative structure, or spatially homogeneous oscillations, depending on the initial condition).

The following system was studied numerically (Bazykin, Markman, 1980):

$$\frac{\partial u}{\partial t} = D_u \frac{\partial^2 u}{\partial z^2} + u^2(1-u) - uv\,,$$
$$\frac{\partial v}{\partial t} = D_v \frac{\partial^2 v}{\partial z^2} - \gamma v(\alpha - u + \delta v) \tag{6.2.1}$$

for $D_u = 0.01$, $D_v = 10$, $\gamma = 1$, $\delta = 0$, and $\alpha = 0$. The diffusion coefficients are chosen to be arbitrary and quite different, while the parameters of the local system are such that the local system is in an oscillatory regime and the bilocal system is in a multistable regime, corresponding to parameter region and phase portrait *4* (Figs. 6.1.2 and 6.1.3).

We have mentioned above, that the unit interval with no influx at the ends is a natural domain of definition of system (6.1.2). There is another ecologically natural domain of definition of the system, which is the unit annulus, that is, we have boundary conditions of the type

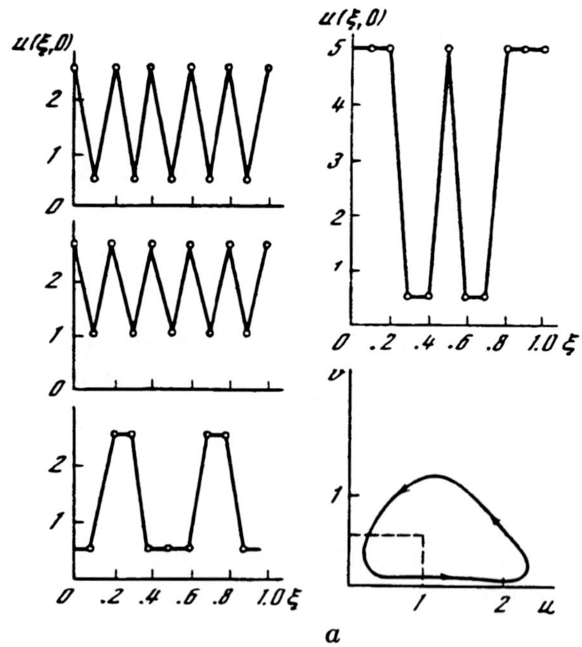

Fig. 6.2.1. Initial distributions of prey population density leading to synchronous homogeneous oscillations all over the annular habitat (a), and to a stationary dissipative structure (b).

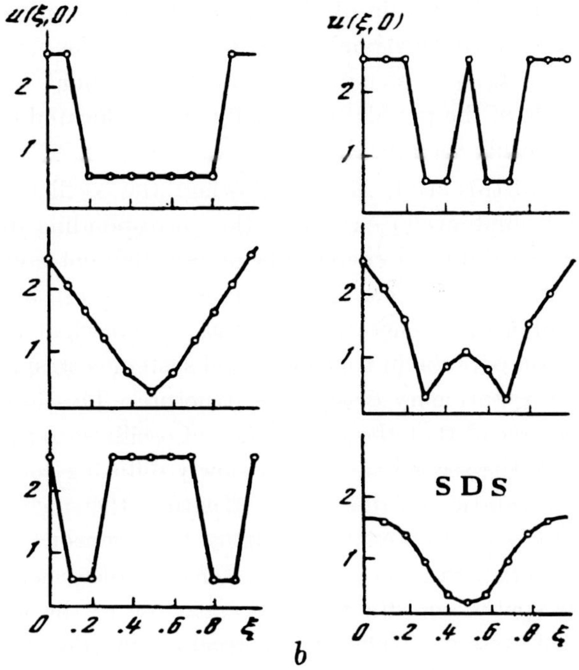

Fig. 6.2.1. (Continued)

$$u(0,\ t) = u(1,\ t);\quad \frac{\partial u}{\partial r}(0,\ t) = \frac{\partial u}{\partial r}(1,\ t)\,;$$
$$v(0,\ t) = v(1,\ t);\quad \frac{\partial v}{\partial r}(0,\ t) = \frac{\partial v}{\partial r}(1,\ t)\,. \tag{6.2.2}$$

Ecologically, (6.2.2) correspond to the distribution of a community over an annular habitat. Such habitats, much different in size, are widespread in nature: banks of ponds and lakes, constant-height levels around mountain ranges and individual mountains, circumpolar regions, etc.

It should be noted that the condition that there is no influx at the ends of the unit interval are similar in meaning to the conditions (6.2.2) of continuity and smoothness on the annulus.

The results of computations by G. S. Markman (Bazykin, Markman, 1980) provided support of the above hypothesis. Figure 6.2.1 shows some of the initial distributions that lead to spatially homogeneous oscillations in densities of populations and to a stable stationary dissipative structure. In all cases the initial distribution $v(r,0)$ was assumed to be homogeneous, such that $v(r,0) \equiv v_0$, where v_0 was taken equal to the unstable stationary value for the solution of the local system (6.1.6). ($v_0 \equiv 0.6$ for all implementations of the model.)

The initial distributions $u(r,0)$ were chosen so that the minimum of the concentration $u(r)$ in the dissipative structure, if it was created, was at $r = 0.5$, and the maximum was at $r = 0$ and $r = 1$, respectively.

We studied the first harmonic of the dissipative structure only, that is, the periodic solution $u(r)$ with only one period on the annulus. For some values of the parameters the dissipative structures of higher harmonics occur, so that two, three or more periods of the population density can be located on a unit annulus. However, we do not study them in this book.

The main result of this study is the following: the oscillatory instability of a local system (stable limit cycle) can drive the corresponding distributed system with strongly differing diffusion coefficients of the different components both to an oscillatory unstable regime (like that of the local system), and to the diffusion unstable regime involving the generation of a stationary dissipative structure. It can be said that a solution periodic in time of a local system corresponds to a spatially periodic solution of a stationary dissipative structure. This interpretation seems to be justified, if we recall that the amplitudes of oscillations in time, in the first case, and in space, in the second case (for a slowly diffusing component), are very close. In the case of a stationary dissipative structure, the spatial distribution of a rapidly diffusing component (a predator) is almost homogeneous, that is, it is almost constant all over the area. In this case, the system's stable density, which is almost independent of the spatial coordinate density, is close to the unstable stationary density for the local system. Thus, on the level of diffusion destabilization of a homogeneous spatial distribution of a slowly diffusing component, one can talk about diffusion stabilization of a locally unstable density for a rapidly diffusing component.

6.3. Evolutionary Appearance of Dissipative Structures

In this section we analyse some possible behaviors of ecological communities that appear as the result of natural Darwinian evolution of populations constituting the community. Of the greatest interest here is the dissipative structure appearing in a bilocal predator–prey system as the result of the survival of a new predominant variety (a *mutant*) in the population.

Before getting to the description of the main content of this section we should recall the following. There are two independent sciences in which mathematical methods were applied earlier and are now being used to study populations and communities: population genetics and mathematical ecology. In population genetics the population is traditionally regarded as being genetically and phenotypically diverse by nature. A priori, some factors of evolution are given: mutation rates, recombination levels, direction and intensity of selection, etc. The question is in which direction and at what rate evolution develops, and what the consequences are. The population is regarded as if it were removed from an ecological community of which it is a part. On the other hand, in mathematical ecology it is traditionally assumed that populations consist of genetically and phenotypically identical species and that the dynamics of those populations within a community should be studied. In this case, the evolution of populations is beyond the scope.

Modern theoretical biology has adopted the point of view that the evolution of a population is determined, first of all, by the selection factors appearing in the process of its interaction with other populations of the community. From this perspective the actual communities are the co-evolution products of the populations involved.

Therefore, the current situation in mathematical and theoretical biology necessitates a synthesis of population-genetic and ecological approaches. In particular, such a synthesis may involve the following. Let us consider a community of populations consisting of genetically and phenotypically different species that interact according to the known laws. We study, first, the directions of selection within a population created by interpopulation interactions, and second, the consequences for the functioning of the community of populations to which evolution leads. This study attempts to make a first step toward combining population-genetic and ecological ideas. We do not concentrate on purely genetic aspects of the evolution, but the species constituting a population will be regarded as genetically isolated, analogous to clones in asexual populations. (The inclusion of the effect of sexual reproduction does not qualitatively affect the results obtained here.)

Note that some of our previous models can be interpreted in terms of evolution of populations and communities. For example, the one-predator–two-preys system (5.2.1) can be regarded as a model of a community consisting of predator and prey populations, the prey population being of two varieties. If, for example, these two varieties are only distinguished by their reproduction coefficients, then in the process of evolution, the variety with the larger coefficient ousts the other. This also has an ecological consequence: the average predator population density decreases due to the evolution of the prey population. More interestingly, the average prey population density remains unchanged even when the evolution process is completed.

The two-predators–one-prey system (5.2.4) admits an analogous interpretation. It is common to these simple models that the qualitative nature of the community dynamics does not change in the process of evolution. However, the ousting of the initial variety (for instance, of a predator) by a mutant, may affect the qualitative behavior of the community while it is in process. In particular, it may lead to the supersession of a stable coexistence by oscillations (Rosenzweig, 1973). Consider a very simple example. Let us return to the predator–prey system (3.4.3) (where we take into account the nonlinear reproduction of prey at small population densities and competition among prey). For simplicity, we consider the limit case $N \gg K$. Then system (3.4.3) is asymptotically approximated by

$$\begin{aligned} \dot{x} &= a'x^2 \frac{K-x}{K} - bxy \,, \\ \dot{y} &= -cy + dxy \,. \end{aligned} \tag{6.3.1}$$

By setting $t = \tau/a$, $x = Ku$, and $y = (a/b)v$ one gets the form

$$\dot{u} = u^2(1-u) - uv\,,$$
$$\dot{v} = \gamma v(\alpha - u)\,, \tag{6.3.2}$$

where $\gamma = dK/a'$ and $\alpha = c/dK$.

The structure of the phase portrait is completely determined by the sign of $\alpha - 1/2$ (Fig. 6.1.1).

Consider now the two-predators–one-prey system, assuming that the predators are only distinguished by the respective value of α:

$$\dot{u} = u^2(1-u) - u(v_{\rm I} + v_{\rm II})\,,$$
$$\dot{v}_{\rm I} = v_{\rm I}(u - \alpha_{\rm I})\,, \tag{6.3.3}$$
$$\dot{v}_{\rm II} = v_{\rm II}(u - \alpha_{\rm II})\,.$$

Setting

$$w = v_{\rm I} + v_{\rm II}, \quad z = v_{\rm I}/v_{\rm II}$$

changes (6.3.3) to

$$\dot{u} = u^2(1-u) - uw\,,$$
$$\dot{w} = w(w - \alpha) - \frac{w}{z+1}\Delta\alpha\,, \tag{6.3.4}$$
$$\dot{z} = -\Delta\alpha z\,,$$

where $\alpha = \alpha_{\rm I}$ and $\Delta\alpha = \alpha_{\rm I} - \alpha_{\rm II}$.

From system (6.3.4) it follows that the invariant coordinate plane $\{v_2 = 0\}$ is repelling, and that $\{v_I = 0\}$ is attracting for $\alpha_{\rm I} > \alpha_{\rm II}$ in the initial $(u, v_{\rm I}, v_{\rm II})$-space of system (6.3.3). In ecological terms, this implies that the predator with the smaller α is more adapted, and as $t \to \infty$, it completely ousts its competitor. Since $\alpha = c/dK$, the predator with a smaller mortality and/or more success in hunting proves to be more adapted. This agrees with the intuitive concept of adaptability. Thus, a natural selection in the predator population included in the predator–prey community leads to a decrease in α.

How does this change the dynamics of the predator–prey system? If $1 > \alpha_{\rm I} > \alpha_{\rm II} > 1/2$, then the extinction of the first predator variety entails only a decrease in the predator equilibrium density. More interesting events occur when $1 > \alpha_{\rm I} > 1/2 > \alpha_{\rm II}$. Assume that initially the predator population consists only of the first variety characterized by $1 > \alpha_{\rm I} > 1/2$, and that the predator and prey populations are in a regime of stable coexistence. Then, as the result of "mutation", there appears a certain small concentration of predators of the second variety, which is characterized by a value of the parameter $\alpha_{\rm II}$ smaller than $1/2$ (for instance due to lower mortality). Then, in due course, the predator of the mutant variety ousts its competitor. The other variety of the predator can no longer coexist in a stable manner with the prey population at constant equilibrium population densities. As a result, the community develops stable oscillations of the predator and prey population densities (Fig. 6.3.1). If $1/2 - \alpha_{\rm II} \ll 1/2$, then the amplitude of these oscillations is small, and one can speak of gradual evolutionary excitation of oscillations.

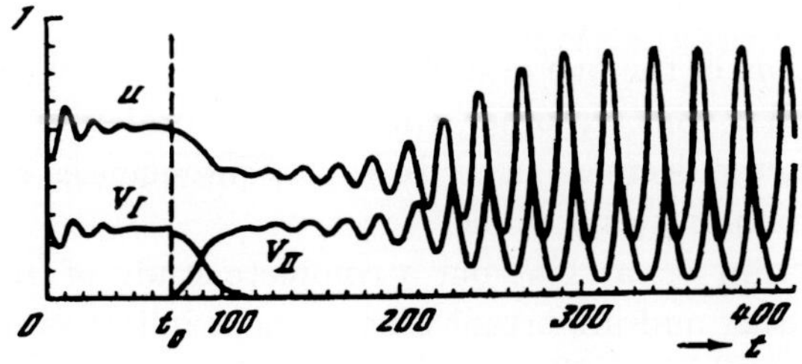

Fig. 6.3.1. Oscillations due to predator evolution in system (6.3.3) for $\alpha_I = 0, \alpha_{II} = 0.435$. From the initial state $u = 0.5, v_I = 0.3, v_{II} = 0$ the system comes to the equilibrium state $\bar{u} = 0.6, \bar{v}_1 = 0.24, \bar{v}_2 = 0$. For t_0, due to "mutation", the system is driven to the state $u = 0.6, v_I = 0.23, v_{II} = 0.01$. The second variety of the predator ousts the first, and oscillations are excited in the system.

Thus, the above example illustrates the situation when the Darwinian evolution of one of the populations of the community leads to a change in the functioning of the community as a whole (the equilibrium is replaced by oscillations). A gradual evolutionary excitation of oscillations in the predator–prey system should not be thought of as the only possible rearrangement in the community in the process of Darwinian evolution of one of its populations. Let us consider a more interesting example of this.

Suppose that the bilocal predator–prey community described by system (6.1.7) has two varieties of the predator with different values of α. Then the system of differential equations takes the form

$$\begin{aligned}
\dot{u}_1 &= u_1^2(1-u_1) - u_1(v_I + v_{II}) - m(u_1 - u_2)\,,\\
\dot{u}_2 &= u_2^2(1-u_2) - u_2(v_I + v_{II}) - m(u_2 - u_1)\\
\dot{v}_I &= v_I\left(\frac{u_1+u_2}{2} - \alpha_I - \delta(v_I + v_{II})\right)\,,\\
\dot{v}_{II} &= v_{II}\left(\frac{u_1+u_2}{2} - \alpha_{II} - \delta(v_I + v_{II})\right)\,.
\end{aligned} \tag{6.3.5}$$

Let us study these equations for $\delta < 1$.

Note that the Arabic and Roman indices used in the system have quite different meanings. The Arabic indices number two spatially separated and ecologically identical prey subpopulations, while the Roman indices are used for two ecologically different varieties within a single predator population.

By using (6.3.4) we obtain the system

$$\begin{aligned}
\dot{u}_1 &= u_1^2(1-u_1) - u_1(v_I + v_{II}) - m(u_1 - u_2)\,,\\
\dot{u}_2 &= u_2^2(1-u_2) - u_2(v_I + v_{II}) - m(u_2 - u_1)\,,
\end{aligned}$$

$$\dot{w} = w_{\rm I}\left[\frac{u_1+u_2}{2} - \alpha_{\rm I} - \delta(v_{\rm I}+v_{II})\right] + \frac{w}{z+1}\Delta\alpha\,, \qquad (6.3.6)$$

$$\dot{z} = -\Delta\alpha z\,.$$

In this situation, as in the one described by (6.3.3), it is obviously the predator with the smaller α that wins. In other words, the natural selection in the predator population results in the decrease in α. To what consequences can this lead for the dynamics of the community as a whole?

Since we do not aim at carrying out a complete study of (6.3.6) we only analyse the most interesting and important case of those initial values of α and δ that provide for only one stable equilibrium in the "ecologically homogeneous" community described by system (6.1.7). That is, the parameters lie in parameter region *1* (Fig. 6.1.2). In other words, the initial value of α is such that the coexistence of predator and prey is stable, and the equilibrium densities of the prey populations are similar in both local subpopulations. For these values of the parameters the bilocal system does not differ from a local one.

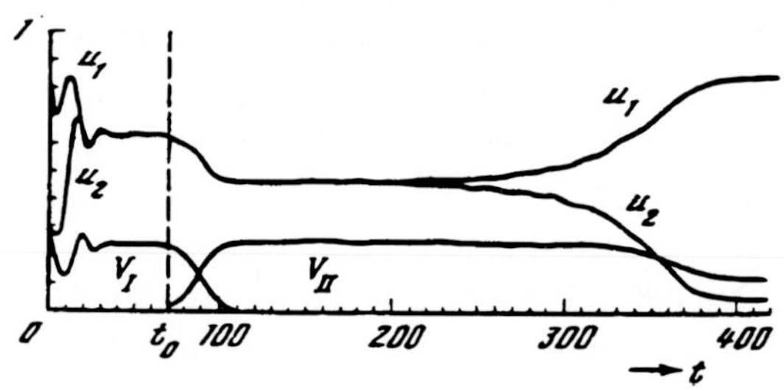

Fig. 6.3.2. Appearance of dissipative structure due to evolution of the predator in system (6.3.5) for $\delta = 0.1$, $m = 0.005$, $\alpha_{\rm I} = 0.6, \alpha_{\rm II} = 0.435$. From the initial state $u_1 = 0.8, u_2 = 0.3, v_{\rm I} = 0.3, v_{\rm II} = 0$ the system moves to the equilibrium state. For t_0, due to "mutation", the system changes to the state $v_{\rm I} = 0.23, v_{\rm II} = 0.01$. The second variety of the predator ousts the first, and a "prey" dissipative structure appears.

Let us now assume that a small concentration of the new, second variety with a smaller value of α appears in the predator population as the result of mutation. This may lead to one of the two, qualitatively different results.

If the migration between the prey subpopulations is intense, so that $m > m_{\rm cr}$ (Fig. 6.1.2), then with a decrease in α the system reaches parameter region *2*. This region corresponds to the globally attracting oscillatory state in which the densities of both prey subpopulations oscillate synchronously and identically. In other words, for a large enough rate of migration between the prey subpopulations the evolution of the predator leads to gradual excitation of oscillations (exact analogues of oscillations in the local systems). Intuitively, this is obvious: for sufficiently intense migration, two linked populations behave as a single one.

The system exhibits a qualitatively different behavior if the migration rate is sufficiently small ($m < m_{\rm cr}$). Suppose that for the first, initial variety of the predator the value of α lies, as before, in parameter region *1*, while for the second (mutant) variety the value of α lies in region *6*. Since this region corresponds to smaller values of α, the mutant variety ousts the initial one. Note that for the initial variety of the predator the only stable behavior is coexistence where the densities of both prey subpopulations are equal. However, for the mutant variety of the predator (region *6*) the stable behaviour is characterized by prey subpopulations with different densities (Fig. 6.3.2).

This means that the ousting of the initial predator by the mutant leads to a nonhomogeneous stationary regime, that is, to a dissipative structure. Note that, due to symmetry, the system can "fall" into one of the equivalent equilibria in the process of evolution. It is difficult to predict which equilibrium that will be.

Let us now notice an interesting peculiarity of the system we studied. For parameters from region *6* near the boundary with region *1* (that is, for not very "small" values of the migration coefficient m) the equilibrium densities of the first and second prey subpopulations do not differ much. Here we deal with a bilocal system in a stationary homogeneous regime that, in the process of evolution, gradually generates a stable dissipative structure. For parameters from region *6* near the boundary with region *8* the prey population densities in the equilibria B_1 and B_2 (dissipative structure) differ a lot. Therefore, the values of the parameter α for the initial and for the mutant predator varieties, lying on opposite sides of the boundary, correspond to what we call the hard evolutionary birth of dissipative structure.

We have presented models illustrating the role of natural selection (within one of the populations constituting a community) on global functional rearrangements of the community as a whole. As a result of Darwinian evolution, a community that is stationary in time and homogeneous in space can become both, oscillatory in time and spatially homogeneous, as well as stationary in time and spatially nonhomogeneous.

Appendix

A 6.1.
Phase portraits
of Eq. 6.1.7

$$\frac{dx}{dt} = x^2(1-x) - xy - m(x-y)$$

$$\frac{dy}{dt} = y^2(1-y) - yz - m(y-x)$$

$$\frac{dz}{dt} = z\left(\frac{x+y}{2} - \alpha - \delta z\right)$$

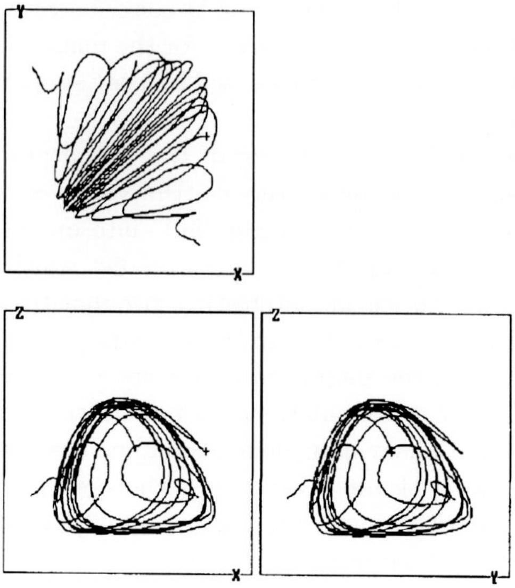

Region 4

α=0.4
δ=0.1
m=0.01

Bibliography

[1] Albrecht, F., Gatzke, H., Haddad, A., Wax, N. (1974) The dynamics of two interacting populations. *J. Math. Anal. and Appl.* **46**, 658–670

[2] Albrecht, F., Gatzke, H., Wax, N. (1973) Stable limit cycles in prey–predator populations. *Science* **181**, 1073–1074

[3] Alekseev, V. V. (1973) Effect of saturation factor on the dynamics of predator–prey system. *Biofizika* **18**, 922–926

[4] Alekseev, V. V. (1975) On the applicability of statistical mechanics methods to describe biocenoses. *Biofizika* **20**, 1133–1136

[5] Alekseev, V. V. (1976) Biogeocenoses–autogenerators and triggers. *Zh. obshchei biologii* **37**, 738–744

[6] Alekseev, V. V., Krymov, I. I., Poliakova, M. S., Sazykina, T. G. (1978) Dynamics and statistical mechanics of biocenoses with biogenic elements of fixed mass. *Chelovek i biosfera* **2**, MGU, Moscow, 42–102

[7] Alekseev, V. I., Polishchuk, E. M., Yusefovich, G. I. (1969) Some mathematical problems of statistical mechanics of biological systems. *Sbornik trudov po agronomicheskoi fizike* **23**, Gidrometeoizdat, Leningrad, 101–127

[8] Allee, W. C., Emerson, A. E., Park, O. (1949) *Principles of animal ecology.* Saunders Co., Philadelphia

[9] Andronov, A. A., Leontovich, E. A., Gordon, I. I., Maier, A. G. (1966) *Qualitative theory of second-order dynamic systems.* Moscow, Nauka; transl. as: Andronov, A. A., Leontovich, E. A., Gordon, I. I., Maier, A. G. (1973) *Qualitative theory of second-order dynamic systems.* Israel Program for Scientific Translations, Jerusalem and John Wiley & Sons, New York

[10] Andronov, A. A., Leontovich, E. A., Gordon, I. I., Maier, A. G. (1967) *Theory of bifurcations of dynamical systems on a plane.* Moscow, Nauka; transl. as: Andronov, A. A., Leontovich, E. A., Gordon, I. I., Maier, A. G. (1971) *Theory of bifurcations of dynamical systems on a plane.* Israel Program for Scientific Translations, Jerusalem

[11] Antonovsky, M. Ya., Fleming, R. A., Kuznetsov, Yu. A., Clark, W. C. (1990) Forest pest interaction dynamics: the simplest mathematical models. *Theor. Pop. Biol.*, **37**, 343–367

[12] Aponina, E. A., Aponin, Yu. M., Bazykin, A. D. (1982) Analysis of complex dynamical behaviour in two prey–one predator model. *Problemy ekologicheskogo monitoringa i modelirovaniya ecosistem* **5**, Gidrometeoizdat, Leningrad, 163–180

[13] Arneodo, A., Coullet, P., Tresser, C. (1980) Occurrence of strange attractors in three-dimensional Volterra equations. *Phys. Lett. A* **79**, 259–263

[14] Arnol'd, V. I. (1978) *Additional chapters of the theory of ordinary differential equations.* Moscow, Nauka; transl as: Arnol'd, V. I. (1983) *Geometrical methods in the theory of ordinary differential equations.* Springer–Verlag

[15] Arnol'd, V. I., Il'yashenko, Yu. S. (1988) Ordinary differential equations. In *Dynamical systems I*, Encyclopaedia of Mathematical Sciences, Springer–Verlag, pp. 3–148

[16] Bazykin, A. D. (1969) A model to describe species numbers dynamics and the problem of coexistence of close species. *Zh. obshchei biologii* **30**, 259–264

[17] Bazykin, A. D. (1974) Volterra system and Michaelis–Menten equation. In *Voprosy matematicheskoi genetiki*, Novosibirsk, pp. 103–143

[18] Bazykin, A. D. (1976) *Structural and dynamical stability of model predator–prey systems.* Laxenburg, IIASA, RM-76-8

[19] Bazykin, A. D. (1978) Principles of systems analysis by the estimation of environmental impacts. *Prirodnye resursy i okruzhayushchaia sreda* **5**, Moscow, 29–41

[20] Bazykin, A. D. (1982) Mathematical modeling of dangerous boundaries in ecology. *Tezisy I Vsesoyuznogo biofizicheskogo s'ezda*, Moscow, pp. 127

[21] Bazykin, A. D., Aponina, E. A. (1981) Model of ecosystem of three trophic levels taking into account the existence of lower critical density of producer population. *Problemy ekologicheskogo monitoringa i modelirovaniya ecosistem* **4**, Gidrometeoizdat, Leningrad, 186–203

[22] Bazykin, A. D., Berezovskaya, F. S. (1979) Allee effect, lower critical population density and dynamics of predator–prey system. *Problemy ekologicheskogo monitoringa i modelirovaniya ecosistem* **2**, Gidrometeoizdat, Leningrad, 161–175

[23] Bazykin, A. D., Berezovskaya, F. S., Buriev, T. I. (1980) Dynamics of predator–prey system including predator saturation and competition. In *Faktory raznoobraziya v matematicheskoi ekologii i populyatsionnoi genetike*, Pushchino, pp. 6–33

[24] Bazykin, A. D., Berezovskaya, F. S., Denisov, G. A., Kuznetsov, Yu. A. (1981) The influence of predator saturation effect and competition among predators on predator–prey system dynamics. *Ecol. Modell.* **14**, 30–57

[25] Bazykin, A. D., Berezovskaya, F. S., Nelgina, O. N., Shvalova, Yu. L. (1981) Lower critical density of predator population and dynamics of predator–prey

system. *Problemy ekologicheskogo monitoringa i modelirovaniya ekosistem* **3**, Gidrometeoizdat, Leningrad, 141–161

[26] Bazykin, A. D., Berezovskaya, F. S., Isaev, A. S., Khlebopros, R. G. (1993) Parametric foundation of stability principle in the dynamics of phytophage–entomophage system. *Dokl. Akad. Nauk, ser. Obshchaya biologiya* **333**, 673–675

[27] Bazykin, A. D., Berezovskaya, F. S. (1995) Mathematical model of dynamics of "phytophage–entomophage" system. In *Problemy monitoringa i modelirovaniya dinamiki lesnykh ekosistem*, ECOLES, Moscow, pp. 309–328

[28] Bazykin, A. D., Buriev, T. I. (1980) Dynamics of predator–prey system by allowance for saturation effects and intraspecific competition. In *Voprosy kachestvennoi teorii differentsial'nykh uravnenii*, SamGU, Samarkand, pp. 31–38

[29] Bazykin, A. D., Buriev, T. I. (1981) Model of dynamics of predator–prey system by allowance for predator saturation, competition among predators for prey, and prey competitions. *Studia biophysica* **83**, 123–130

[30] Bazykin, A. D., Khibnik, A. I. (1979) On the study of a system of differential equations in the neighbourhood of parametric point $(0, \pm i)$. In *5ya Vsesoyuznaya konferentsiya po kachestvennoi teorii differentsial'nykh uravnenii*, Kishinev, pp. 10–11

[31] Bazykin, A. D., Khibnik, A. I. (1981a) On the sharp generation of self-oscillations in a model of Volterra type. *Biofizika* **26**, 851–853

[32] Bazykin, A. D., Khibnik, A. I. (1981b) Soft and sharp generation of oscillations in predator–prey system. In *Matematicheskie modeli kletochnych populyatsiy*, Gorky State University, pp. 53–69

[33] Bazykin, A. D., Khibnik, A. I. (1982) Bilocal model of dissipative structure. *Biofizika* **27**, 132–136

[34] Bazykin, A. D., Khibnik, A. I., Aponina, E. A. (1983) A model of evolutionary appearance of dissipative structure in ecosystems. *J. Math. Biol.* **18**, 13–23

[35] Bazykin, A. D., Khibnik, A. I., Aponina, E. A., Neifel'd, A. A. (1980) A model of evolutionary appearance of dissipative structure in ecological systems. In *Faktory raznoobraziya v matematicheskoi ekologii i populyatsionnoi genetike*, Pushchino, pp. 33–47

[36] Bazykin, A. D., Markman, G. S. (1980) On dissipative structures in ecological systems. In *Faktory raznoobraziya v matematicheskoi ekologii i populyatsionnoi genetike*, Pushchino, pp. 135–149

[37] Bautin, N. N., Leontovich, E. A. (1976) *Methods and techniques of qualitative study of dynamic systems on a plane.* Nauka, Moscow

[38] Bautin, N. N., Shil'nikov, L. P. (1980) Behaviour of dynamic systems near the boundaries of the domain of stability of equilibrium states and periodic motions (dangerous and non-dangerous boundaries). In Marsden, J. and McCracken, M., *Bifurkatsiya rozhdeniya tsykla i ee prilozheniya*, Mir, Moscow, pp. 294–316

[39] Belyakov, L. A. (1981) On the structure of bifurcation manifolds in systems

with separatrix loop of saddle–focus. In *IX Mezhdunarodnaya konferentsiya po nelineinym kolebaniyam*, Kiev, pp. 53–55

[40] Bogdanov, R. I. (1976a) Reducing vector field on a plane to the orbital normal form. *Functsional'nyi analis i ego prilozheniya* **10**, 73–74

[41] Bogdanov, R. I. (1976b) Versal unfolding of singular point of vector field on a plane in case of zero eigenvalues. *Trudy seminara im. Petrovskogo* **2**, MGU, Moscow, 37–65

[42] Brauer, F., Soudack, A. C. (1979) Stability region in predator–prey systems with constant-rate prey harvesting. *J. Math. Biol.* **8**, 55–71

[43] Buriev, T. I. (1983) *Analysis of dynamics of interacting populations under stationary and variable external conditions (mathematical models).* Avtoref. kand. diss., Pushchino

[44] Busenberg, S., Kumar, S. K., Austin, P., Wake, G. (1990) The dynamics of a model of a plankton–nutrient interaction. *Bull. Math. Biol.* **52**, 677–696

[45] Caswell, H. (1972) A simulation study of the time-lag population model. *J. Theor. Biol.* **34**, 419–439

[46] Cramer, N. F., May, R. M. (1972) Interspecific competition, predation and species diversity: A comment. *J. Theor. Biol.* **34**, 289–293

[47] Cohen, D. S., Murray, J. D. (1981) A generalized diffusion model for growth and dispersal in a population. *J. Math. Biol.* **12**, 237–249

[48] Collins, J. B., Wollkind, D. J. (1990) A global analysis of a temperature-dependent model system for a mite predator–prey interaction. *SIAM J. Appl. Math.* **50**, 1348–1372

[49] Collins, J. B., Wollkind, D. J., Moody, M. E. (1990) Outbreaks and oscillations in a temperature-dependent model of a mite predator–prey interaction. *Theoret. Pop. Biol.* **38**, 159–192

[50] Denisov, G. A., Kuznetsov, Yu. A. (1981) Qualitative study of a mathematical model describing the exploitation of animal resources. In *Matematicheskoe modelirovanie agrobiotsenoticheskih protsessov*, Nauka, Moscow, pp. 152–164

[51] Dubois, D. M. (1975) A model of patchiness for prey–predator plankton population. *Ecol. Modell.* **1**, 67–80

[52] Eman, T. I. (1966) On some mathematical models of biogeocenoses. *Problemy kibernetiki* **16**, 19–202

[53] Fife, P. C. (1976) Pattern formation in reacting and diffusing systems. *J. Chem. Phys.* **64**, 554–564

[54] Fife, P. C. (1979) *Mathematical aspects of reacting and diffusing systems.* Lect. Notes in Biomathematics, Springer, New York

[55] Foerster, N. V., Mora, P. M., Amiot, L. W. (1960) Doomsday: Friday, 13, November, A.D. 2026. *Science* **132**, 1291–1295

[56] Forrester, J. (1971) *World dynamics.* Write-Allen Press Inc., Cambridge, Mass

[57] Freidlin, M. I., Svetlosanov, V. A. (1976) On the effect of small random disturbances on the stability of states of ecological systems. *Zh. obshchey biologii* **37**, 715

[58] Freedman, H. J. (1979) Stability analysis of a predator–prey system with mutual interference and density dependent death rates. *Bull. Math. Biol.* **41**, 67–78

[59] Fujii, K. (1979) Complexity-stability relationship of two prey–one predator species system model: Local and global stability. *J. Theor. Biol.* **69**, 613–623

[60] Garfinkel, G., Sack., R. (1964) Digital computer simulation of an ecological system based on modified mass action law. *Ecology* **45**, 502–507

[61] Gause, G. F. (1933) Mathematical approach to the problems of struggle for existence. *Zoologicheskii Zh.* **12**, 170–177

[62] Gause, G. F. (1934) *The struggle for existence.* Williams & Wilkins, Baltimore

[63] Gause, G. F., Vitt, A. A. (1934) On the periodical oscillations of population sizes. Mathematical theory of relaxational interactions between predators and preys and its application to a population of two Protozoa. *Izv. Akad. Nauk, series 7*, 1551–1559

[64] Gilpin, M. E. (1972) Enriched predator–prey systems: theoretical stability. *Science* **177**, 902–904

[65] Gilpin, M. E. (1979) Spiral chaos in a predator–prey model. *Amer. Natur.* **113**, 306–308

[66] Gorban', A. N., Okhotin, V. I., Sadovsky, M. G., Khlebopros, R. G. (1982) *A simplest equation of mathematical ecology.* Preprint, Inst. of Forestry and Wood, Siberian Branch of the USSR Academy of Sciences, Krasnoyarsk

[67] Gorstko, A. B. (1976) Simulation system "The Azov Sea". *Trudy VNIRO* **118**, 48–55

[68] Hainzl, J. (1988) Stability and Hopf bifurcation in a predator–prey system with several parameters. *SIAM J. Appl. Math.* **47**, 170–190

[69] Haken, H. (1978) *Synergetics.* Springer-Verlag

[70] Harrison, G. W. (1986) Multiple stable equilibria in a predator–prey system. *Bull. Math. Biol.* **48**, 137–148

[71] Hastings, A. (1981) Multiple limit cycles in a predator–prey model. *J. Math. Biol.* **11**, 51–63

[72] Hilborn, R. (1979) Some longterm dynamics of predator–prey models with diffusion. *Ecol. Modell.* **6**, 23–30

[73] Hirata, H. A. (1980) A model of hierarchical ecosystems with migration. *Bull. Math. Biol.* **42**, 119–130

[74] Holling, C. S. (1965) The functional response of predator to prey density and its role in mimicry and population regulation. *Mem. Entomol. Soc. Canada* **45**, 1–60

[75] Holling, C. S. (1973) Resilience and stability of ecological systems. *Ann. Rev. Ecol. Syst.* **4**, 1–23

[76] Holling, C. S. (ed.) (1978) *Adaptive environmental assessment and management.* John Wiley & Sons

[77] Hsu, S. B. (1978a) On the global stability of a predator–prey system. *Math. Biosci.* **39**, 1–10

[78] Hsu, S. B., Hubbel, S. P., Waltman, P. (1978b) Competing predators. *SIAM J. Appl. Math.* **35**, 617-625

[79] Hsu, S. B. (1981) Predator mediated coexistence and extinction. *Math. Biosci.* **54**, 231–248

[80] Hsu, S. B., Hubbel, S. P., Waltman, P. (1978) Contribution to the theory of competing predators. *Ecol. Monogr.* **48**, 337–349

[81] Huberman, C. (1978) Qualitative behaviour of fishery system. *Math. Biosci.* **42**, 1–14

[82] Hutchinson, G. E. (1948) Circular causal systems in ecology. *Ann. N.Y. Acad. Sci.* **50**, 221–246

[83] Isaev, A. S., Nedorezov, L. V., Khlebopros, R. G. (1979a) *The "boomerang effect" in the models of separate populations and elementary biocenoses.* Preprint, Institute of Forestry and Wood, Siberian Branch of USSR Academy of Sciences, Krasnoyarsk

[84] Isaev, A. S., Nedorezov, L. V., Khlebopros, R. G. (1979b) A mathematical model of the escaping effect in predator–prey interaction. In *Matematicheskii analiz komponentov lesnykh biogeotsenozov*, Nauka, Novosibirsk, pp. 74–82

[85] Isaev, A. S., Khlebopros, R. G., Nedorezov, L. V. (1979c) *Qualitative analysis of the phenomenological model describing the dynamics of population numbers of forest insects.* Preprint, Institute of Forestry and Wood, Siberian Branch of the USSR Academy of Sciences, Krasnoyarsk

[86] Ivlev, V. S. (1955) *Experimental ecology of fish feeding.* Pishchepromizdat, Moscow

[87] Izrael', Yu. A. (1976) Permissible anthropogenic load on the environment. In *Vsestoronnii analiz okruzhaiushchei prirodnoi sredy* , Proceedings of the 2nd Soviet–American Symposium, Gidrometeoizdat, Leningrad, pp. 12–20

[88] Izrael', Yu. A. (1977) A concept of biosphere monitoring. In *Monitoring sostoyaniya okruzhaiushchei sredy*, Gidrometeoizdat, Leningrad, pp. 10–26

[89] Izrael', Yu. A., Filippova, L. M., Insarov, G. E., Semevskii, F. N., Semenov, S. M. (1981) Ecological monitoring and control of the natural environment. *Problemy ecologicheskogo monitoringa i modelirovaniya ekosistem* **4**, Gidrometeoizdat, Leningrad, 6–19

[90] Kasarinoff, N., Deiesch, P. van der (1978) A model of predator–prey system with functional response. *Math. Biosci.* **39** , 124–134

[91] Kerner, E. H. (1955) A statistical mechanics of interacting biological species. *Bull. Math. Biophys.* **19**, 121–146

[92] Kerner, E. H. (1957) Further consideration of the statistics of biological associations. *Bull. Math. Biophys.* **21**, 217–255

[93] Khibnik, A. I. (1989) LINLBF: A program for continuation and bifurcation analysis of equilibria up to codimension three. In *Continuation and Bifurcations: Numerical Techniques and Applications*, Kluwer Academic Publishers, pp. 283–296

[94] Khibnik, A. I. (1990) *Using TRAX: Tutorial to accompany TRAX, a program for the simulation and analysis of dynamical systems.* Exeter Sowtware, Setauket, New York

[95] Khibnik, A. I., Kuznetsov, Yu. A., Levitin, V. V., Nikolaev, E. V. (1993) Continuation techniques and interactive software for bifurcation analysis of ODEs and iterated maps. *Physica* **D62**, 360–371

[96] Kirlinger, G. (1988) Permanence of some ecological systems with several predator and one prey species. *J. Math. Biol.* **26**, 217–232

[97] Koch, A. L. (1974) Competitive coexistence of two predators utilizing the same prey under constant environmental conditions. *J. Theor. Biol.* **44**, 373–386

[98] Kolesov, Yu. S. (1979) Mathematical models in ecology. In *Issledovaniya po ustoichivosti i teorii kolebanii*, Yaroslavl', pp. 3–40

[99] Kolesov, Yu. S., Shvitra, D. I. (1979a) Role of lag in mathematical models. *Litovskii matematicheskii sbornik* **19**, 115–128

[100] Kolesov, Yu. S., Shvitra, D. I. (1979b) *Self-oscillations in systems with a lag.* Vilnius

[101] Kolmogoroff, A. (1936) Sulla theoria di Volterra della lotta per l'essistenza. *G. Inst. Ital. degli attuari* **7**, 74–80

[102] Kolmogorov, A. N. (1972) Qualitative study of mathematical models of population dynamics. *Problemy Kibernetiki* **26**, Moscow, Nauka, 100–106

[103] Kolmogorov, A. N., Petrovsky, I. G., Piskunov, N. S. (1937) A study of equation of diffusion associated with the growth of substance, and its application to a certain biological problem. *Bull. MGU, Ser. matematika i mekhanika* **1**, 26

[104] Leslie, P. H. (1945) On the use of matrices in certain population mathematics. *Biometrika* **33**, 183–212

[105] Leslie, P. H. (1948) Some further notes of the use of matrices in population mathematics. *Biometrika* **35**, 213–245

[106] Levin, S. A. (1980) Non-uniform stable solutions to reaction–diffusion equations: Applications to ecological pattern formation. In *Proc. Intern. Sympos. on Synergetics*, Springer

[107] Levin, S. A., Segal, S. A. (1976) An hypothesis to explain the origin of planktonic patchiness. *Nature* **259**, 659

[108] Levitin, V. V. (1989) *TRAX: Simulation and analysis of dynamical systems.* Exeter Software, Setauket, New York

[109] Lotka, A. J. (1925) *Elements of physical biology.* Williams & Wilkins, Baltimore

[110] Lotka, A. J. (1956) *Elements of mathematical biology.* Dover, New York

[111] Lukyanov, V. I. (1982) On bifurcations of dynamical systems with loop of the "saddle–node" separatrix. *Differentsial'nyye Uravneniya* **18**, 1493–1506

[112] Lyapunov, A. A. (1972) On cybernetic problems of biology. *Problemy Kybernetiki* **25**, Nauka, Moscow, 5–39

[113] Lyapunov, A. A., Bagrinovskaya, G. P. (1975) On methodological questions of mathematical biology. In *Matematicheskoe modelirovanie v biologii*, Nauka, Moscow, pp. 5–18

[114] MacFadyen, A. (1963) *Animal ecology: aims and methods.* Pitman, New York

[115] Malthus, T. R. (1798) *An essay on the principle of population.* Johnson, London [Cambridge University Press, New York, 1992]

[116] Masayasu, M. (1979) Asymptotic behaviors of parabolic system related to a plankton prey and predator model. *SIAM J. Appl. Math.* **37**, 499–512

[117] May, R. M. (1972) Limit cycles in predator–prey communities. *Science* **177**, 900–902

[118] May, R. M. (1973) Comment to Albrecht's *et al.* paper. *Science* **181**, 1074

[119] May, R. M. (1974) *Stability and complexity in model ecosystems.* Princeton University Press

[120] May, R. M. (1975) Biological populations obeying difference equations: stable points, stable cycles and chaos. *J. Theor. Biol.* **51**, 511–524

[121] Menshutkin, V. V. (1971) *Mathematical modeling of populations and communities of water animals.* Nauka, Leningrad

[122] Mimura, M., Murray, I. D. (1978) On a diffusive predator–prey model which exhibits patchiness. *J. Theor. Biol.* **75**, 249–252

[123] Moiseev, N. N. (1979) *Mathematics performs an experiment.* Nauka, Moscow

[124] Molchanov, A. M. (1975a) Mathematical models in ecology. Role of critical regimes. In *Matematicheskoe modelirovanie v biologii*, Nauka, Moscow, pp. 131–141

[125] Molchanov, A. M. (1975b) Critical points of biological systems (mathematical models). In *Matematicheskoe modelirovanie v biologii*, Nauka, Moscow, pp. 142–153

[126] Molchanov, A. M., Nazarenko, V. G., Shaturnyi, I. G. (1971) Biophysics of complex systems: mathematical models. *Biofizika* **16**

[127] Molchanov, A. M., Bazykin, A. D. (1979) Method of system analysis and mathematical modeling in solving the problems of the program "Man and Biosphere" in the USSR. In *Proc. of the 1st Conference of National Committee of Socialist Countries on the Program MAB*, Moscow, pp. 223–226

[128] Monod, L. (1942) *Recherches sur la croissance des cultures bacteriennes.* Herman, Paris

[129] Moran, P. A. P. (1962) *Statistical processes of evolution theory.* Oxford University Press

[130] Nakajima, H., De Angelis, D. L. (1989) Resilience and local stability in a nutrient limited resource–consumer system. *Bull. Math. Biol.* **51**, 501

[131] Nedorezov, L. V. (1979) *Dynamics of outbreaks of mass propagation of forest insect phytophagans (mathematical model).* Avtoref. kand. diss., Krasnoyarsk

[132] Nicholson, A. I. (1954) An outline of the dynamics of animal populations. *Austral. J. Zool.* **2**, 9–65

[133] Nozdracheva, V. P. (1982) Bifurcation of structurally unstable separatrix loop. *Differentsial'nyye Uravneniya* **18**, 1551–1558

[134] Odum, E. (1971) *Fundamentals of ecology (3rd ed.)*. Saunders, Philadelphia

[135] Okhonin, V. A. (1979) Kinetic equations of populations. In *Matematicheskii analiz komponentov lesnykh biogeotsenozov*, Nauka, Moscow, pp. 119–131

[136] Okhonin, V. A., Khlebopros, R. G. (1978) *Kinetic equations of the state of populations*. Preprint, Institute of Forestry and Wood, Siberian Branch of the USSR Academy of sciences, Krasnoyarsk

[137] Paine, R. T. (1966) Food web complexity and species diversity. *Amer. Natur.* **100**, 65–75

[138] Parish, J. D., Saila, S. B. (1970) Interspecific competition: Predator and species diversity. *J. Theor. Biol.* **27**, 207–220

[139] Pearl, R. (1927) The growth of populations. *Quart. Rev. Biol.* **2**, 532

[140] Pearl, R. (1925) *The biology of population growth*. A.A. Knopf, New York

[141] Polishchuk, E. M. *et al.* (1969) Some mathematical problems of statistical mechanics of biological systems. *Sbornik trudov po agronomicheskoi fizike* **23**, Gidrometeoizdat, Leningrad, 101–127

[142] Polishchuk, E. M. (1971) On some statistical criteria of stability of multicomponent biocenoses. *Sbornik trudov po agronomicheskoi fizike* **30**, Gidrometeoizdat, Leningrad, 202–211

[143] Poluektov, R. A. (ed.) (1974) *Dynamical theory of biological populations*. Nauka, Moscow

[144] Polyakova, M. S., Sazykina, T. G. (1976) Structure and stability of model biogeocenoses consisted of six species. *Zh. obshchei biologii* **37**, 745–751

[145] Pykh, Yu. A. (1983) *Equilibrium and stability in models of population dynamics*. Nauka, Moscow

[146] Razzhevaikin, V. N. (1981a) On the appearance of dissipative structures in a "predator–prey" type system. In *Avtovolnovye protsessy v sistemakh s diffusiei*, Gorky, pp. 243–249

[147] Razzhevaikin, V. N. (1981b) Spatially distributed two-level ecosystems. In *Matemat. metody v ekologii i genetike*, Nauka, Moscow, pp. 36–50

[148] Ricker, W. E. (1954) Stock and recruitment. *J. Fish. Res. Board Canada* **11**, 559–663

[149] Romanovsky, Yu. M., Stepanova, N. V., Chernavsky, D. S. (1975) *Mathematical modeling in biophysics*. Nauka, Moscow

[150] Rosenzweig, M. L. (1973) Evolution of the predator isocline. *Evolution* **27**, 84–94

[151] Rosenzweig, M. L., MacArthur, R. H. (1963) Graphical representation and stability conditions of predator–prey interactions. *Amer. Natur.* **97**, 209–223

[152] Rössler, O. E. (1976) Different types of chaos in two simple differential equations. *Zs. Naturforsch.* **31**, 1664–1670

[153] Segal, L. A., Jackson, J. (1972) Dissipative structure: An explanation and an ecological example. *J. Theor. Biol.* **37**, 545–559

[154] Segal, L. A., Levin, S. A. (1973) Application of nonlinear stability theory to studies of the effects of diffusion on predator–prey interactions. *Proc. AJP Conf.* **27**, 123–152

[155] Sidorin, A. P. (1981) Behaviour of populations with several stationary states in a random environment. In *Matematicheskie modeli v ekologii i genetike*, Nauka, Moscow, pp. 71–74

[156] Shapiro, A. P. (1974) Mathematical models of competition. *Upravlenie i informatsiya* **10**, Vladivostok, 5–75

[157] Shapiro, A. P., Luppov, S. P. (1983) *Recurrent equations in the theory of population biology.* Nauka, Moscow

[158] Sharkovsky, A. N. (1964) Existence of cycles of continuous transformation of straight line into itself. *Ukrainskii matematicheskii zh.* **16**, 61–65

[159] Shil'nikov, L. P. (1963) On some random births of periodic motions from singular trajectories. *Matematicheskii sbornik* **61**, 443–446

[160] Shil'nikov, L. P. (1965) On a certain case of existence of a countable set of periodic motions. *Dokl. Akad. Nauk SSSR* **160**, 558–561

[161] Shil'nikov, L. P. (1970) On the problem concerning structure of neighbourhood of structural stable equilibrium of saddle–focus type. *Matematicheskii sbornik* **81**, 92–103

[162] Shil'nikov, L. P. (1980) Theory of bifurcations and Lorentz model. In Marsden, J. and McKracen, M., *Bifurkatsiya rozhdeniya tsykla i ee prilozheniya*, Mir, Moscow, pp. 317–336

[163] Shimazu, Y., Sugiyama, K., Kojima, T., Tomida, E. (1972) Some problems in ecology oriented environmentology. II. Terrestrial environmentology. *J. Earth. Sci. Nagoya Univ.* **20**, 31–89

[164] Shklovsky, I. S. (1965) *Universe, life, intelligence* Nauka, Moscow

[165] Skaletskaya, E. I., Frisman, E. Ya., Shapiro, A. P. (1979) *Discrete models of population dynamics and harvesting optimization.* Nauka, Moscow

[166] Smith, J. M. (1974) *Models in ecology.* Cambridge University Press

[167] Smith, H. L. (1982) The interaction of steady-state and Hopf bifurcations in a two predator–one prey competition model. *SIAM J. Appl. Math.* **42**, 27–43

[168] Steel, J. H. (1974a) Spatial heterogeneity and population stability. *Nature* **83**, 83

[169] Steel, J. H. (1974b) Stability of plankton ecosystems. In *Ecological Stability*, Chapman and Hall, London, pp. 179–191

[170] Sullivan, P. J. (1988) Effect of boundary, region length, and diffusion rates on a specially heterogeneous predator–prey. *Ecol. Modell.* **43**, 225

[171] Svirezhev, Yu. M. (1976) Vito Volterra and modern mathematical ecology. Afterward, in V. Volterra, *Matemat. teoriia bor'by za sushchestvovanie*, Nauka, Moscow, pp. 245–283

[172] Svirezhev, Yu. M., Logofet, D. O. (1978) *Stability of biological communities.* Nauka, Moscow

[173] Takahashi, F. (1964) Reproduction curve with two equilibrium points: a consideration on the fluctuation of insect population. *Res. Pop. Ecol.* **6**, 28–36

[174] Turing, A. M. (1952) The chemical basis of morphogenesis. *Philos. Trans. Roy. Soc. London B* **237**, 37–72

[175] Vance, R. R. (1978) Predation and resource partitioning in one predator– two prey model communities. *Amer. Natur.* **119**, 797–813

[176] Vasil'ev, V. A. (1976) Dynamic dissipative structures. In *Termodinamika biologicheskikh protsessov*, Nauka, Moscow, pp. 197–210

[177] Verhulst, P. F. (1838) Notice sur la loi que la population suit dans son accroissement. *Corr. Math. et Phys.* **10**, 113–121

[178] Volterra, V. (1926) Variazone e fluttuazini del numero d'individui in specie animali conviventi. *Mem. Accad. naz. Lincei. Ser. 6*, **2**, 31–113

[179] Volterra, V. (1931) *Lecons sur la theorie mathematique de la lutte pour la vie.* Gauthiers-Villars, Paris

[180] Wangersky, P. I., Cunningham, W. J. (1957) Time lag in prey–predator population models. *Ecology* **3**, 136–139

[181] Watt, K. (1968) *Ecology and resource management.* McGraw-Hill, New York

[182] Wollkind, D. J., Collings, J. B., Logan, J. A. (1988) Multistability in a temperature-dependent model system for prey–predator mite outbreak interactions of the fruit trees. *Bull. Math. Biol.* **50**, 389

[183] Wollkind, D. J., Collings, J. B., Barba, M. C. B. (1991) Diffusive instabilities in a one-dimensional temperature-dependent model system for a mite predator–prey interaction on a fruit trees: Dispersal motility and aggregative prey taxis effects. *J. Math. Biol.* **29**, 339–362

[184] Zaikin, A. N., Rudyakov, Yu. A. (1989) Distribution criterion for the pelagic community and kinetic regimes of its euristic models. *Dokl. Akad. Nauk SSSR* **306**, 977–981

[185] Zhdanov, Yu. A., Gorstko, A. B. (1975) Mathematical model of rational utilization of water resources of the Azov basin. *Vodnye resursy*, No. 3, 188–192